SpringerBriefs in Environmental Science

SpringerBriefs in Environmental Science present concise summaries of cutting-edge research and practical applications across a wide spectrum of environmental fields, with fast turnaround time to publication. Featuring compact volumes of 50 to 125 pages, the series covers a range of content from professional to academic. Monographs of new material are considered for the SpringerBriefs in Environmental Science series.

Typical topics might include: a timely report of state-of-the-art analytical techniques, a bridge between new research results, as published in journal articles and a contextual literature review, a snapshot of a hot or emerging topic, an in-depth case study or technical example, a presentation of core concepts that students must understand in order to make independent contributions, best practices or protocols to be followed, a series of short case studies/debates highlighting a specific angle.

SpringerBriefs in Environmental Science allow authors to present their ideas and readers to absorb them with minimal time investment. Both solicited and unsolicited manuscripts are considered for publication.

More information about this series at http://www.springer.com/series/8868

Santosh Kumar Sarkar

Loricate Ciliate Tintinnids in a Tropical Mangrove Wetland

Diversity, Distribution and Impact of Climate Change

 Springer

Santosh Kumar Sarkar
Department of Marine Science
University of Calcutta
Calcutta
India

ISSN 2191-5547 ISSN 2191-5555 (electronic)
ISBN 978-3-319-12792-7 ISBN 978-3-319-12793-4 (eBook)
DOI 10.1007/978-3-319-12793-4

Library of Congress Control Number: 2014953312

Springer Cham Heidelberg New York Dordrecht London

Printed on acid-free paper

Springer is part of Springer Science+Business Media (www.springer.com)

Preface

This concise text first provides a detailed case study highlighting the diversity, distribution, species identification and impact of natural and anthropogenic stresses of a specialized microzooplankton group—loricate ciliate Tintinnids—in an iconic tropical mangrove wetland—Sundarbans, the tide-dominated, vulnerable and prograding megadelta. The field of this choreotrich ciliate biodiversity is rapidly developing as new research challenges and getting intense interest due to its key position in marine planktonic food web as major consumers of pico- and nanoplankton. Hence, they play a crucial role in transferring element and energy from lower small protists to larger protists and mesozooplankton and also constitute an important element of global change research. There is complete dearth of information related to regional case studies of this specialized group of microzooplankton in Sundarbans wetland environment which acts as potential refuge of living marine resources. Hence, the author has taken keen interest in proposing this book based on systematic survey and collections of microzooplankton in coastal regions of Sundarbans during 2010–2012, acclaimed as the World Heritage tropical site in Asia and also designated as a global biodiversity 'hotspot'. The book gains critical importance as these low-lying coastal zones (LLCZ) are experiencing wide range of pressures due to siltation, eutrophication, coastal development, aquaculture and climate change. These are also considered as disaster 'hot spots' potentially vulnerable to extreme events impacted by slow onset processes that include sea level rise, biodiversity loss, flooding, severe cyclones, etc., which act as the key stressors on pelagic organisms including plankton.

The book is intended to serve as a reliable and up-to-date reference source for students, researchers and teachers, who are actively involved in the field of protozoology/marine plankton ecology/marine biology. This would no doubt also serve as a valuable source of useful information for policy makers, coastal zone managers and large sections of people engaged in coastal research and development.

Acknowledgements

The author is highly grateful to Dr. Wuchang Zhang (Key Laboratory of Marine Ecology and Environmental Sciences, Institute of Oceanology, Chinese Academy of Sciences, China), Dr. S. Barria (Instituto Argentino de Oceanografía, Argentina) for identification of Tintinnida. I wish to express my sincere thanks and gratitude to Dr. J.R. Dolan (Marine Microbial Ecology, Centre National de la Recherche Scientifique (CNRS), France for identification of tintinnids as well as for the microphotographs used in the book. We wish to thank Dr. Zhang for critically examining a part of the manuscript and giving his valuable comments and suggestions for improving the text matter.

The author gratefully acknowledge the support from Dr. R.K. Sarangi, Scientist, SAC (ISRO), Ahmedabad, India for collaborative work and study on the phenomenon of algal bloom in Sundarbans wetland ecosystems by NASA/TMI and MODIS-AQUA satellite data.

The dynamic assistance rendered by my research scholar Shri Dibyendu Rakshit as well as my daughter Ms. Prathama Sarkar for preparing the manuscript is greatly appreciated. My research scholars and other scientists have extended support in one form or the other and I would like to thank them collectively.

The author is grateful to the Editor of the journal CLEAN–Soil, Air and Water Pollution, Wiley-VCH Verlag, Germany for reproducing the research article titled "Occurrence of the centric diatom *Hemisdiscus hardmannianus* (Bacillariophyceae) and its impact on water quality characteristics and plankton community structure in Indian Sundarbans mangrove wetland", 2013, 41(4): 333–339. The publication of the book would not have been possible without the generous funding jointly from the Council of Scientific and Industrial Research (CSIR) (Sanction No. 23(0014)/09/EMR-II) and University Grants Commission, New Delhi, India (Sanction No. 40-388/2011 (SR)).

Finally, I would like to express our sincere gratitude to Alexandrine Cheronet, Senior Publishing Editor, and Judith Terpos, Senior Editorial Assistant, Environmental Sciences, Springer for her keen interest in the publication of the book. Thanks are also due to anonymous reviewers for their valuable and constructive suggestions and comments which helped in giving the book its final shape.

The ecological importance of microzooplankton (MZP) in the pelagic food webs of the world's oceans has long been recognized. These organisms often constitute a significant component of the plankton community in many marine environments. Tintinnids are planktonic choreotrich ciliates forming the most distinctive and dominant microzooplankton group, characterized by their possession of a tubular or vase-shaped shell or lorica whose architecture forms the basis of classic taxonomic schemes. Lorica morphology is not only a valuable taxonomic characteristic but has also been linked to ecological characteristics of tintinnids, especially in terms of feeding activity.

Contents

Chapter 1
Introduction

Microzooplankton (MZP) are defined as heterotrophic organisms with body dimensions between 20 and 200 μm in size and can be divided into four broad groups: (i) ciliates dominated by Spirotrichea subclasses, Oligotrichia and Choreotrichia; (ii) dinoflagellates and other large flagellates; (iii) rotifers; and (iv) larvae of microcrustaceans dominated by copepod nauplii. These taxonomically diverse communities constitute a significant component of the plankton community in many marine environments (Relevante et al. 1987; Burkill et al. 1987; Pierce and Turner 1994). By virtue of their small body size, these organisms can exploit small food particles unavailable to most meso- and macrozooplankton, and thus act as trophic intermediaries between pico- and nanoplankton and meso- and macrocarnivores (Stoecker and Capuzzo 1990; Gifford 1991; Roman and Gauzens 1997; Klein Breteler et al. 1999). In recent years, there has been an upsurge of interest in research on MZP in the marine ecosystem owing to their higher mass-specific physiological rates compared with larger zooplankton.

Tintinnids are the best known group among protistian microplankton, characterized by their possession of a tubular or vase-shaped shell or lorica whose architecture forms the basis of classic taxonomic schemes. Its morphology is not only a valuable taxonomic characteristic but has also been linked to ecological characteristics of tintinnids especially in terms of feeding activity. The diameter of the mouth end of the lorica, the lorica oral diameter (LOD), is related to the size of the food items ingested by the ciliate. Heinbokel (1978) noted that the largest prey ingested were about half the LOD, and Dolan et al. (2002) found that tintinnid feeding rates were maximal on prey sizes equal to about 25 % of LOD. Thus, the lorica distinguishes species both taxonomically and ecologically in tintinnid ciliates.

They are ubiquitous in marine as well as in estuarine waters, and they are distributed globally into five categories, namely, cosmopolitan, neritic, warm-temperate, boreal, and austral. They can easily be detected since species identification is based upon morphological characteristics of preservable loricae (Dolan and Gallegos 2001). Individual species (untreated or preserved) may be identified by examination using a transmitted light microscope rather than requiring cytological staining, biochemical screening, or RNA sequencing. Chemical preservation method may destroy the fragile ciliate cell, leaving the hard lorica unaffected.

© The Author(s) 2015
S.K. Sarkar, *Loricate Ciliate Tintinnids in a Tropical Mangrove Wetland*,
SpringerBriefs in Environmental Science, DOI 10.1007/978-3-319-12793-4_1

Tintinnids can be divided into two groups corresponding to distinguished lorica types corresponding to different habitats, either coastal or open sea waters: (i) Agglutinated (or agglomerated) lorica composed of mineral particles and generally encountered in the genera *Tintinnopsis, Stenosemella, Tintinnidium,* and *Leprotintinnus* inhabiting in the coastal waters (ii) Hyaline lorica which lacks extraneous particles in the lorica. Although some hyaline lorica species are found in coastal regions (e.g., *Favella*), they are dominant forms in open sea water tintinnid assemblage.

The importance of ciliates was initially associated mainly with the microbial loop and corresponding microbial web, but now there is increasing evidence that these protists are also a crucial part of the herbivorous web, consuming a wide spectrum of particle sizes from bacteria to large diatoms and dinoflagellates as well as other ciliates. As a consequence, in the past decades, much research effort has been devoted to finding factors affecting ciliate abundance and distribution and their trophic behavior in different environments. Despite the increasing importance of MZP as trophic intermediaries in the pelagic as well as marine food web, ocean energy flow and carbon cycling, ecological significance, population dynamics, and trophodynamic roles of MZP remain poorly investigated in tropical waters, particularly in Indian coastal marine and estuarine waters. Currently, the knowledge of MZP is increasing rapidly due to the recognition of their dynamic role in marine coastal systems. Hence, studies on MZP have been well recognized and included as an important component in the Global Ocean Ecosystem Dynamics (GLOBEC) and Joint Global Ocean Flux Study (JGOFS) programmes where they have been shown to be the major movers of energy, carbon cycling, and remineralization of nutrients in the oceans.

Research on MZP will emerge as one of the important thrust areas in biological oceanography, especially in studies on the microbial food web, where MZP is acting as a central link between the bacterial population and metazoan consumers and larval fish consumers. Furthermore, MZP communities are the food of some fish larvae; hence, they have lucrative values in shellfish and finfish culture, especially first feeding larval fishes (Stoecker and Capuzzo 1990; van der Meeren and Naess 1993). Interestingly, in Antarctic pelagic zone, food chain was traditionally believed to be simple starting with diatoms that are grazed by krill and consumed by carnivores (El-Sayed 1971).

1.1 Microzooplankton and Tintinnid

MZP are very specialized group of plankton community inhabiting in the coastal or marine waters with definite community structure. They constitute a considerable portion of the zooplankton biomass in marine and estuarine environments (Porter et al. 1985; Pierce and Turner 1992). Due to their ubiquitous in distribution, small size, and rapid metabolic growth rates (Heinbokel 1978; Verity 1985; Fenchel 1987), much attention has been given to their role as primary consumers of pico- and nano-sized autotrophs and heterotrophs, as well as nutrient regenerators and as important food sources for metazoan zooplankton and fish larvae (Dolan et al. 2002; Pierce and Turner 1992). They also represent one of the most morphologically diverse groups in the plankton community (Bachy et al. 2012). Furthermore, they often represent an important link between the microbial fraction and the larger grazers (Laval-Peuto et al. 1986; Pierce and Turner 1994). Due to small body size, MZP have higher weight- specific physiological rates such as feeding, respiration, excretion, and growth than larger metazoans (Fenchel 1987; Verity 1985). Thus, MZP acts as a trophic intermediate between pico/nanoplankton and larger zooplankton.

Ciliates of the MZP include tintinnid ciliates, a suborder of choreotrich ciliates characterized by the possession of a species-specific external shell or lorica composed of protein, shaped like a bowl or vase or tube, within which the ciliate cell can withdraw. Tintinnids are large subgroup (>1,000 species) of marine ciliates feeding primarily on nanoplanktonic diatom and photosynthetic flagellates. Although tintinnids are nearly always a minority component of the ciliate community, they are ideal for studies of species distributions, diversity, and changes in the structure or composition of MZP communities, because changes in the composition of their populations can easily be detected since species identification is based upon morphological characteristics of preservable loricae (Dolan and Gallegos 2001). Although it is accepted that tintinnid taxonomy should be complemented with cytological and molecular analyses, such information is currently available for fewer than 30 species (Agatha and Stüder-Kypke 2007). This represents 5 % of the total species included in the most w-idely used classification system, which encompasses hundreds of taxa often distinguished on the basis of variations in the shape, size, and details of the lorica (Kofoid and Campbell 1929, 1939). Therefore, lorica morphology remains the most important feature in tintinnid taxonomy (Santoferrara and Alder 2009; Kršinić 2010a). Thus, individuals, untreated other than preserved, may be identified by examination using a transmitted light microscope rather than requiring cytological staining, biochemical screening, or RNA sequencing.

On the other hand, due to their hard loricae, tintinnids are the best known group of marine ciliates, but the durability of loricae can also cause problems. Collection with plankton nets has been found to cause tintinnids to abandon loricae (Paranjape and Gold 1982), and preservation may also destroy the fragile ciliate cell, leaving the lorica unaffected. Empty loricae sink at rates of up to 1.5 md^{-1} (Smayda and Bienfang 1983), and loricae of estuarine tintinnids can be carried hundreds of kilometers offshore before settling to the sediments (Echols and Fowler 1973). Thus, it is difficult to tell if empty loricae in samples were occupied at the time of collection or were abandoned and transported to different areas by currents.

Lorica morphology is not only a valuable taxonomic characteristic but has also been linked to ecological characteristics of tintinnids especially in terms of feeding activity (Dolan 2010). The maximal prey size ingested has been reported to be about 45 % of the LOD and preferred prey size about 25 % of LOD (Heinbokel 1978; Dolan et al. 2002).

1.2 Why Tintinnid Has Been Given Priority?

Tintinnids, as loricate choreotrich ciliates, constitute a single suborder of the order Choreotrichidae and are thus phylogenetically coherent group (e.g., Lynn 2008) of morphologically and ecologically similar species (e.g., Dolan 2010). In terms of morphology, all are characterized by the possession of a shell (lorica) whose architecture forms the basis of classic traditional taxonomic schemes. About 1,200 species have been described (Agatha and Strüder-Kypke 2012), virtually all restricted to the marine plankton. The diameter of the mouth end of the lorica, LOD, is a conservative taxonomic character (Laval-Peuto and Brownlee 1986). Because LOD is a conservative taxonomic character, the morphological diversity of an assemblage, in terms of different sizes of LODs, is closely related to taxonomic diversity (Dolan 2006). From paleontological point of view, tintinnids could be considered as typically one of the oldest among contemporary taxa of the plankton.

Unlike the numerically dominant taxa of the MZP, the oligotrichic ciliates and heterotrophic dinoflagellates, tintinnids have species-specific shells or loricae. Hence, species identifications are relatively straightforward, unlike most microplankton taxa. Generally, the tintinnid lorica is tubular or conical, but a very large range of constructions is found among the different taxa. The diameter of the open mouth end of the lorica, the LOD, is an important characteristic not only in terms of taxonomy but it is also a key ecological characteristic. A given species feeds at a maximum rate on prey of a size of about 25 % of LOD (Dolan 2010). In common with other protistan taxa with skeletal structures or shells, the characteristics of such hard parts are used to define species despite the fact that some tintinnid species are known to display considerable plasticity in lorica morphology (e.g., Laval-Peuto and Brownlee 1986).

Compared with other planktonic protists, tintinnids have some specific advantages. They are more abundant than most other shelled MZP (e.g., foraminifers, radiolarians), have higher duplication rates, and inhabit both coastal and open-ocean waters. On the other hand, unlike the phytoplanktons, which are dependent on light, tintinnids have subsurface and deep-dwelling species useful for tracing undercurrents and upwellings. From the ecological point of view, MZP in general, and tintinnids in particular, can play a major role in pelagic communities. Tintinnid grazing impact on the phytoplankton may exceed that of larger consumers several fold, utilizing up to 100 % of the primary production (Capriulo and Carpenter 1983). In addition, they often represent an important link between the microbial fraction and the larger grazers (Laval-Peuto and Brownlee 1986; Pierce and Turner 1994).

While the tintinnid ciliate component of MZP communities may play a relatively minor role in processes such as carbon flux or nutrient regeneration, they are ideal organisms for the study of changes in the structure or composition of MZP communities (Thompson et al. 1999). This is because changes in the composition of tintinnid communities can easily be detected as species identifications can be made using lorica morphology. Tinitinnids can be identified more conveniently by examining specimens in plankton settling chambers; in contrast, identification of naked oligotrichs demands more technical processes like mounting on slides, cytological staining, and examination under high magnification (i.e., protargol silver staining; Montagnes and Lynn 1987, 1991).

References

Agatha, S., & Strüder-Kypke, M. C. (2012). Reconciling cladistics and genetic analyses in choreotrichid ciliates (Protists, Spirotrichea, Oligotrichea). *Journal of Eukaryotic Microbiology, 59*, 325–350.

Agatha, S., & Stüder-Kypke, M. C. (2007). Phylogeny of the order Choreotrichida (Ciloiophora, Spirotricha, Oligotrichea) as inferred from morphology ultrastructure, ontogenesis, and SSr RNA gene sequences. *European Journal of Protistology, 43*, 37–63.

Bachy, C., Gómez, F., López-García, P., Dolan, J. R., Moreira, D. (2012). Molecular phylogeny of tintinnid ciliates (Tintinnida, Ciliophora). *Protist, 163*, 873–887.

Burkill, P. H., Mantoura, R. F. C., Llewellyn, C. A., Owens, N. J. P. (1987). Microzooplankton grazing and selectivity of phytoplankton in coastal waters. *Marine Biology, 93*, 581–590.

Capriulo, G. M., & Carpenter, E. J. (1983). Abundance, species composition and feeding impact of tintinnid microzooplankton in Central Long Island Sound. *Marine Ecology Progress Series, 10*, 277–288.

Dolan, J. R., & Gallegos, C. L. (2001). Estuarine diversity of tintinnids (planktonic ciliates). *Journal of Plankton Research, 23*, 1009–1027.

Dolan, J. R., Claustre, H., Carlotti, F., Plounevez, S., Moutin, T. (2002). Microzooplankton Diversity: Relationship of tintinnid ciliate with resources, competitors and predators from the Atlantic coast of Morocco to the eastern Mediterranean. *Deep Sea Research Part I, 49*, 1217–1232.

Dolan, J. R. (2006). Re-discovered beauty new images for old descriptions tropical tintinnids of the genus *Xystonellopsis* Ciliophora, Tintinnia. *Protist, 157*, 251–253.

Dolan, J. R. (2010). Morphology and ecology in tintinnid ciliates of the marine Plankton: Correlates of lorica dimensions. *Acta Protozoologica, 49*, 235–244.

Echols, R. J., & Flower, G. A. (1973). Agglutinated tintinnid loricae from some recent and late Pleistocene shelf sediments. *Micropalaentology, 19*, 431–443.

El-Sayed, S. Z. (1971). Biological aspects of the pack-ice ecosystem. In G. Deacon (Ed.), *symposium on Antarctic ice and water masses* (pp. 35–54). Tokyo: Scientific Communications on Antarctic Research.

Fenchel, T. (1987). *Ecology of Protozoa—The Biology of Free Living Phagotrophic Protists* (p. 197). Berlin: Springer-Verlag.

Gifford, D. J. (1991). The protozoan-metazoan trophic link in pelagic ecosystems. *Journal of Protozoology, 38*, 81–86.

Heinbokel, J. F. (1978). Studies on the functional role of tintinnids in the Southern California Bight: Grazing and growth rates in laboratory cultures. *Marine Biology, 47*, 177–189.

Klein Breteler, W. C. M., Schogt, N., Baas, M., Schouten, S., Kraay, G. W. (1999). Trophic upgrading of food quality by proto zoans enhancing copepod growth: Role of essential lipids. *Marine Biology, 135*, 191–198.

Kofoid, C. A., & Campbell, A. S. (1929). A conspectus of the marine and freshwater Ciliata belonging to the sub-order Tintinnoinea with descriptions of new species principally from the Agassiz Expedition to the eastern tropical Pacific, 1904–1905. *University of California Publications in Zoology, 34*, 1–403.

Kofoid, C. A., & Campbell, A. S. (1939). Reports on the scientific results of the expedition to the Eastern tropical Pacific. The Ciliata: The Tintinnoinea. *Bulletin of the Museum of Comparative Zoology, 84*, 1–473.

Kršinić F. (2010a) Tintinnids (Tintinnida, Choreotrichia, Ciliata) in the Adriatic Sea, Mediterranean. Part I. Taxonomy. Split, Croatia: Institute of Oceanography and Fisheries; Dalmacija papir, Split kurzer Diagnose derneuen Arten. Ergebn. Atlant. Ozean Planktonexped. Humboldt-Stift, *3*, 1–33.

Laval-Peuto, M., & Brownlee, D. C. (1986). Identification and systematics of the Tintinnina (Ciliophora): Evaluation and suggestions for improvement. In *Annales de l'Institut océanographique* (Vol. 62, No. 1, pp. 69–84). Institut océanographique.

Laval-Peuto, M., Heinbokel, J. F., Anderson, O. R., Rassoulzadegan, F., Sherr, B. F. (1986). Role of micro- and nanozooplankton in marine food webs. *Insect Science Application*, *7*, 387–395.

Lynn, D. H. (2008). *The ciliated protozoa: characterization, classification, and guide to the literature* (3rd ed.). New York: Springer.

van der Meeren, T., & Naess, T. (1993). How does cod (*Gadus morhua*) cope with the variability in feeding conditions during early larval stages? *Marine biology*, *112*, 637–647.

Montagnes, D. W. S., & Lynn, D. H. (1987). A quantitative protargol stain (QPS) for ciliates: A description of the method and tests of its quantitative nature. *Marine Microbial Food Webs, 2*, 83–93.

Montagnes, D. J. S., & Lynn, D. H. (1991). Taxonomy of choreotrichs, the major marine planktonic ciliates, with emphasis on the aloricate forms. *Marine Microbial Food Webs, 5*, 59–74.

Paranjape, M. A., & Gold, K. (1982). Cultivation of marine pelagic protozoa. *Annales de l'Institut océanographique*, *58*(S), 143–150.

Pierce, R. W., & Turner, J. T. (1992). Ecology of planktonic ciliates in marine food webs. *Reviews in Aquatic Sciences*, *6*, 139–181.

Pierce, R. W., & Turner, J. T. (1994). Plankton studies in Buzzards Bay, Massachusetts, USA: IV. Tintinnids, 1987 to 1988. *Marine Ecology Progress Series, 112*, 235–240.

Porter, K. G., Sherr, E. B., Sherr, B. F., Pace, M., Sanders, R. W. (1985). Protozoa in planktonic food webs. *Journal of Protozoology*, *32*, 409–415.

Relevante, N., Gilmartin, M., Smodlaka, N. (1987). The effects of Po River induced eutrophication on the distribution and community structure of ciliated protozoan and micrometazoan populations in the northern Adriatic Sea. *Journal of Plankton Research*, *7*, 461–471.

Roman, M. R., & Gauzens, A. L. (1997). Copepod grazing in the equatorial Pacific. *Limnologica Oceanography*, *42*, 623–634.

Santoferrara, L. F., & Alder, V. A. (2009). Morphological variability, spatial distribution and abundance of *Helicostomella* species (Ciliophora: Tintinnina) in relation to environmental factors (Argentine shelf; 40–558S). *Marine Science*, *73*, 701–716.

Smayda, T. J., & Bienfang, P. K. (1983). Suspension properties of various phyletic groups of phytoplankton and tintinnids in an oligotrophic subtropical system. *Marine Ecology*, *4*(4), 289–400.

Stoecker, D. K., & Capuzzo, J. M. (1990). Predation on protozoa: Its importance to zooplankton. *Journal of Plankton Research*, *12*, 891–908.

Thompson, G. A., Alder, V. A., Boltovskoy, D., & Brandini, F. (1999). Abundance and biogeography of tintinnids (Ciliophora) and associated microzooplankton in the Southwestern Atlantic Ocean. *Journal of Plankton Research, 21*, 1265–1298.

Verity, P. G. (1985). Grazing, respiration, excretion and growth rates of tintinnids. *Limnologie Oceanography*, *30*, 1268–1282.

Chapter 2
Methodology

Abstract The Indian Sundarbans wetland, the largest delta situated at the mouth of the Ganges River Estuary (GRE), is situated in the low-lying, meso–macrotidal, humid and tropical belt, which harbors the world's largest mangrove forest together with associated flora and fauna. The total estuarine phase of the wetland is very irregular and criss-crossed by several tributary rivers, creeks, and waterways. Five study sites of different hydrodynamic conditions were chosen covering the eastern and western flank of Sundarbans coastal region. These sites belong to a lower deltaic plain experiencing intense semidiurnal tides and wave action with a meso–macrotidal setting (3–6 m amplitude). Detailed of the selection of study sites, sampling strategy, and numerical analyses of the water quality and plankton are discussed.

Keywords Community structure · Water quality · Lorica · Tintinnids morphology

2.1 Study Sites

India has a large coastline and vast stretches of coastal wetlands, estuaries, bays, backwater lagoons, and mangrove swamps, which extend over large part of the coasts. The term "tide-dominated wetland" has been used by Selvan (2003) for the Indian Sundarbans, as the study area involves the tide-dominated estuarine stretch of the Hugli (Ganges) and Matla rivers, the intertidal creeks and canals and the marshy mangrove swamps of Sundarbans. The wetland (7,658 km^2) is characterized by a large wealth of water resources, faunal and floral biodiversity, commercially exploitable species of shell and fin fishes (Bhattacharya and Sarkar 2003), and recreational potential and luxuriant wildlife including a tiger reserve (Naskar and Guha Bakshi 1987; Stanley and Hait 2000). Shared between two neighboring countries, Bangladesh and India, the larger part (62 %) is situated in the southwest corner of Bangladesh. At present, the total land area is 4,143 km^2 (including exposed sandbars of 42 km^2), and the remaining water area of 1,874 km^2 encompasses rivers, small streams, and canals. As it is the transition zone of freshwater and saltwater, it offers a specialized habitat for both plant and animals. A significant

© The Author(s) 2015

S.K. Sarkar, *Loricate Ciliate Tintinnids in a Tropical Mangrove Wetland*,
SpringerBriefs in Environmental Science, DOI 10.1007/978-3-319-12793-4_2

part of this wetland has been undergoing denudation due to deforestation, agriculture, and reclamation of lands.

The Indian Sundarbans (21° 32′ N–22° 40′ E and 88° 85′ N–89° 00′ E), the largest delta in the estuarine phase of the River Ganges, is famous for its luxuriant mangrove vegetation, acclaimed as UNESCO World Heritage Site for its capacity of sustaining rich and diverse fauna and floral communities. Situated in the low-lying, meso–macrotidal, humid and tropical belt, the area is interspersed with a large number of islands (~101 in number) and tidal channel systems through which semidiurnal tides of meso–macrotidal amplitude interplay with moderate to strong wave effects. Due to combination of interaction of geography, topography, and climate, the delta has a variety of habitats and ecosystems, inhabited by globally threatened species, critically endangered and endangered species.

The total estuarine phase of the Indian Sundarbans is very irregular and crisscrossed by several tributary rivers (such as Matla, Gosaba, Muriganga, Saptamukhi), creeks, and waterways. The area is characterized by shallow waters which are constantly mixed by wind and tidal currents. The velocity of the current varies considerably with the state of the tides and the season in this estuarine system. Both the flood and ebb currents go up to 6 knots during high spring tides. This highly turbid estuary allows a scarce light penetration due to a great amount of organic and inorganic suspended materials. Like other estuaries, this is also characterized by a predominant monsoon regime and the total 12-month period in a year was classified into three different seasons, namely premonsoon (March–June) dry season with occasionally higher temperature, monsoon (July–October) accompanied by heavy rainfall (annual average precipitation is ~1,800 mm) and postmonsoon (November–February) characterized by lower temperatures and lower precipitations. Seven sampling sites [Dhamakhali (S_1), Canning (S_2), Basanti (S_3), Lot 8 (S_4), Phuldubi (S_5), Chemaguri (S_6), and Gangasagar (S_7)] have been selected covering both eastern and western flank of Sundarbans (as shown in Fig. 2.1). They are of diverse environmental stresses and of different hydrodynamic conditions in the context of depth, tidal amplitude, and wave action gradually being less toward the upstream direction (Table 2.1).

2.2 Sampling Strategy and Analytical Protocol

2.2.1 Collection and Preservation of Tintinnid

Collection of tintinnids For tintinnids, 1,000 ml of surface water samples were collected by precleaned plastic bottles from each site and immediately preserved with Lugol's solution (2 % final concentration, volume/volume) and stored refrigerated in darkness except during transport and settling (Dolan 2002).

Settlement of the sample In the laboratory, the water samples were placed in measuring cylinders of 1,000 ml with 2 special outlets at the level of 500 and

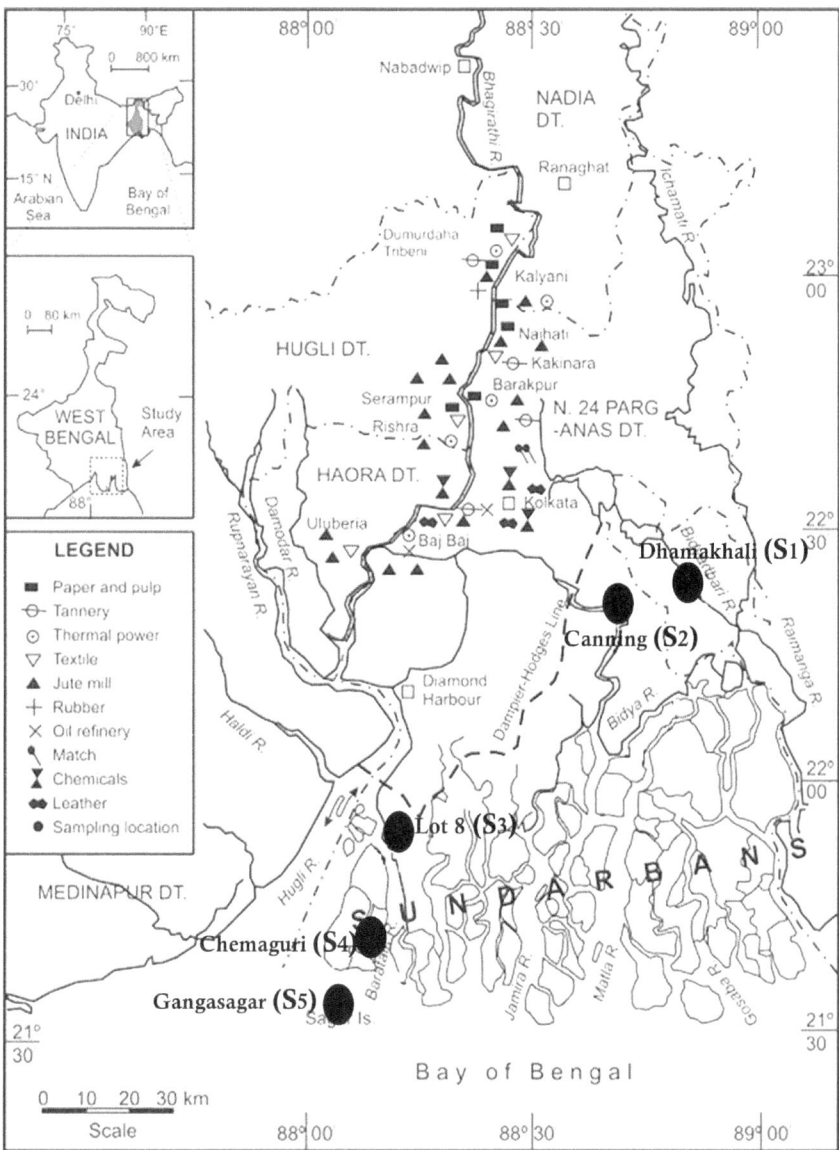

Fig. 2.1 Map of Indian Sundarbans showing the location of five study sites (S_1–S_5). The location of multifarious industries is shown in the upstream of Hugli River along with the primary rivers and tributaries

250 ml, respectively, which were blocked by clumps. After 24 h, when almost all the planktons were settled at the bottom of the cylinder, the clumps were being opened and the water from the upper portion of the measuring tube was allowed to flow out without disturbing the last 250 ml sample. After that, the sample was

Table 2.1 Justification for the selection of the study sites in Sundarbans coastal regions

Characteristic features of the study sites	Dhamakhali (River Vidyadhori) S_1	Canning (River Matla) S_2	Lot 8 (River Muriganga) S_3	Chemaguri (narrow creek) S_4	Gangasagar (mouth of the estuary) S_5
Geomorphic coordinates	22° 20′ 26″ N 88° 53′ 18″ E	22° 12′ 13″ N 88° 40′ 01″ E	21° 52′ 34″ N 88° 09′ 06″ E	21° 40′ 45″ N 88° 09′ 03″ E	21° 36′ 10″ N 88° 01′ 43″ E
Tidal environment	Semidiurnal micro (<2 m)	Semidiurnal meso (2–4 m)	Semidiurnal macro (>4 m)	Semidiurnal macro (>4 m)	Semidiurnal macro (>4)
Wave action	Low	Low	Moderately low	Moderately low	High
Distance from sea (km)	100.53	94.71	29.73	5	0
Depth (m)	~2.0–3.0	~2.0–3.5	~2.0	~3.0–4.6	~8.0–9.8
Anthropogenic stress	Boating, use of antifouling paints, domestic sewage	Agriculture fishing, boating	Boating, agricultural runoff, domestic sewage	Boating, fishing, antifouling paints, ice factory discharge	Fisheries, boating, antifouling paints

transferred to a beaker and again kept for 24 h. Lastly, the sample was concentrated to a volume of 25 ml by siphoning out the rest of the volume from the beaker (Godhantaraman 2002).

Qualitative analysis From the concentrated volume (25 ml) of the sample, it was taken on a glass slide with a dropper for identification and observed under a phase-contrast microscope at a magnification of 40×. As there are no other criteria for identification, tintinnids were identified using lorica morphology (Lorica shape, size as well as the agglomeration type) described by Kofoid and Campbell (1929, 1939) and Marshall (1969).

Quantitative analysis For quantitative analysis, the sample was taken as a drop of known volume on a glass slide with the help of a micropipette. Three to five aliquots of each sample were counted, and the mean value was considered.

Calculation of lorica volume, biomass, and production rate of tintinnids Biomass was calculated as the total biovolume of the tintinnid community for each sampling date. Both lorica length and oral diameter were measured, and volumes were calculated by assigning standard geometric configurations. Lorica volume (LV, μm^3) was converted to body carbon weight using the regression equation: $444.5 + 0.053$ LV (Verity and Langdon 1984). Production rate (P, $\mu g\ Cl^{-1}\ day^{-1}$) was estimated from biomass (B, $\mu g\ Cl^{-1}$) and empirically determined specific growth rate (g, day^{-1}): $P = B \times G$. Multiple regression, $1.52 \ln T - 0.27 \ln CV - 1.44$, where T is the temperature (°C) and CV is the cell volume (μm^3) proposed by Müller and Geller (1993) for ciliates, was used.

2.2.2 Collection and Analysis of Water Samples

Collection of water samples Simultaneously with plankton samples, surface water samples were also collected for the analyses of temperature, pH, salinity, turbidity, dissolved oxygen (D.O.), biological oxygen demand (B.O.D.), nutrients (nitrate, phosphate and silicate), and chlorophyll pigments. Water samples were taken in precleaned plastic bottles and were immediately preserved in 4 °C and taken to the laboratory for further analyses. For chlorophyll pigment analyses, one liter of surface water sample was wrapped by black paper and was taken to the laboratory for further analyses. For dissolve oxygen, water samples were taken in B.O.D. bottles (125 ml) and the D.O. was fixed immediately using Winkler's reagents.

Temperature For measurement of water temperature, 1 L of seawater on board was taken and a centigrade thermometer of ±0 °C accuracy was dipped into the water immediately after their collections and the water temperature was recorded after 5 min, when the mercury level stood constant. Temperature values are expressed in °C.

pH The pH of the water samples was recorded with the help of a Deluxe Digital pH Meter (Model No. 101 E) considering three replicate samples of each station studied.

Salinity Salinity was measured on board using a refractometer. Immediately after collection of the water sample, the instrument was rinsed with distilled water for 2 or 3 times and it was standardized at zero. Then, sample water was added to it drop by drop and the reading was taken. This process was repeated for 5 times, and the mean reading was taken and expressed in parts per thousand (ppt).

Turbidity Turbidity was measured using a turbidity meter (Model no: SYS 304 E).

Dissolved oxygen The D.O. content was estimated adopting Winkler's titrimetric method. The water samples drawn carefully into 125 ml B.O.D. avoiding the interference of air bubbles were treated with 1 ml each of the Winkler's I and Winkler's II solution. The bottles were restoppered and well shaken. This was done in the filed immediately after the collections are over and was transported to the laboratory for further analysis. The brownish white precipitates of manganous hydroxide were dissolved by adding 1 ml of concentrated H_2SO_4, which liberated iodine equivalent to the original D.O. The brown color solution was qualitatively transferred to a 250 ml conical flask and was titrated against standard sodium thiosulfate solution of 0.01 N (W/V). 2 % starch solution (W/V) was used as the indicator. Change of blue color, developed due to the reaction between iodine and starch, to colorless was taken as the end point. From the amount of thiosulfate consumed in the titration, the D.O. content in each sample was calculated employing the formula below. The values of oxygen are presented as mg l^{-1}.

Calculations

$$\text{Dissolved oxygen in mg } l^{-1} = \frac{CD \times N \times E \times 1,000}{4 \times (V-2)}$$

CD   Burette reading for thiosulfate titration
N   Normality of thiosulfate (0.01 N)
E   Equivalent weight of oxygen (32,000)
V   Volume of the B.O.D. bottle (125 ml)

Biological oxygen demand B.O.D. is the measure of the degradable organic material present in a water sample and can be defined as the amount of oxygen required by the microscopic organisms in stabilizing the biologically degradable organic matter under aerobic condition. The principle of this measuring is the difference of oxygen concentration between the sample and after incubating it after 5 days at 20 °C. Water sample was taken in a precleaned plastic container (>1 l), and it was made oxygen saturated by compressed air bubbling using an aerator for 30 min. After the saturation, 1 l of that water was taken and 1 ml each of the reagents, phosphate buffer, magnesium sulfate solution, calcium chloride solution, and ferric chloride solution were added to it and mixed thoroughly. Two B.O.D. bottles were taken for the analysis, and both the bottles were filled with water and stoppered. D.O. was immediately determined using Winkler's titrimetric method, and the second bottle was incubated at 20 °C for five days. After completion of five days, D.O. of the second bottle was determined using the same procedure. B.O.D. was calculated using the following formula, and B.O.D. was expressed by mg l^{-1}.

Calculations B.O.D. (mg l^{-1}) = D.O. content of the first bottle (mg l^{-1}) − D.O. content of the second bottle after five days (mg l^{-1}).

Estimation of Nitrate–Nitrite All nitrates present in the sample water were converted to nitrite by reduction. A glass column packed with copper-coated cadmium cheeps was used for reduction. Method was based on the formation of azo dye. 100 ml of water sample was mixed with ammonium chloride solution (2 ml of 25 %) and passed through the amalgamated cadmium redactor column with a speed of 2 drops s − 1. The effluent (50 ml) collected from the column was then treated with 1 ml solution of sulfanilamide; the resultant diazonium ion was coupled with 1 ml of N-(1-Naphthy)ethylenediamine dihydrochloride to give an intensely pink dye. The absorbance of the resulting pink solution was measured spectrophotometrically at 543 nm against a reagent blank. Efficiency of the redactor column (>90 %) was tested periodically with standards and was subjected to identical treatment in each batch. The concentration of total nitrate and nitrite was computed from calibration curve.

Dissolved inorganic Phosphate–Phosphorus Phosphate–phosphorus of sample water was determined using acidified molybdate solution and ascorbic acid. 35 ml of sample water was treated with 1 ml mixed reagent (mixture of 0.073 M ammonium molybdate and 9.1 N H_2SO_4 and a small portion of potassium antimonyl tartrate) followed by the addition of 1 ml 0.4 M ascorbic acid solution. Dissolved inorganic phosphate present in seawater was converted to the formation of phosphomolybdate complex with acidified molybdate reagent, which on reduction with ascorbic acid formed a highly colored molybdenum blue compound. The absorbance of the resultant molybdenum blue was measured spectrophotometrically at 882 nm against reagent blank. Samples and standards were subjected to identical treatment in each batch. Turbidity blank was used whenever it was necessary. The concentration of PO_4^{-3}–P was then computed from calibration curve.

Dissolved Silicate–Silicon Silicate–silicon was determined using acidified molybdate solution and ascorbic acid solution. Oxalic acid was added to avoid interference of phosphate in samples. 25 ml of sample and a set of standards were mixed with 1 ml of mixed reagent (mixture of equal volume of 0.16 M ammonium heptamolybdate solution and 7.3 N H_2SO_4 acid solution) followed by the addition of 1 ml oxalic acid solution (0.7 M) and 1 ml 0.1 M ascorbic acid solution. Determination of dissolved silicate was based on the formation of a yellow silicomolybdic complex when an acidified sample was treated with ammonium molybdate solution. This on reduction with ascorbic acid yields intensely colored molybdenum blue complex. The blue silicomolybdic complex is formed within 30 min and stable for hours. The absorbance of the blue complex was measured photo-metrically at 810 nm against reagent blank, and the concentration of $SiO_4{}^{-4}$–Si was computed from calibration curve.

Estimation of Chlorophyll Pigment Concentration To estimate the chlorophyll pigment (a, b and c) concentration in the surface water of the sampling sites, 1 l of the surface water was collected in a acid-washed dark plastic bottles, kept in ice and taken to the laboratory. In laboratory, the collected water sample was filtered through a Millipore filter paper (0.45 μ) using a suction pump. After the filtration, the fil-ter paper was extracted in a acid-cleaned centrifuge tube by 10 ml 90 % acetone and kept in dark and freezed for 24 h. Next day, the content in the centrifuge tube was homogenized in a acid-cleaned glass homogenizer and again transferred into the centrifuge tube. Then, it was centrifuged at 20,000 rpm for 20 min. The clear solu-tion was decanted from the tube, and the absorbance was measured at 630, 645 and 665 nm. The concentration of the chlorophyll pigments was expressed in mg l^{-1}, and it was formulated using the following formula (Strickland and Parsons 1972).

Calculations

Chlorophyll a (mg pigment/m^3) = 11.6 E665 − 1.31 E645 − 0.14 E630
Chlorophyll b (mg pigment/m^3) = 20.7 E645 − 4.34 E665 − 4.42 E630
Chlorophyll c (mg pigment/m^3) = 55 E630 − 4.64 E665 − 16.3 E645

where E stands for the extinction values at wavelengths indicated by the sub-scripts, measured in 10 cm cells after correcting for a blank.

2.2.3 Statistical Analyses

Species diversity (H'), evenness (J'), and richness (d) of samples were calculated as follows: the equation: $H' = -\sum_{i=1}^{s} Pi(\ln Pi)$, $J' = H/\ln(S)$, and $d = (s - 1)/\ln(N)$, where Pi = proportion of the total count arising from the ith species; S = total no of species; and N = total no of individuals (Xu et al. 2008). Correlations were cal-culated using Pearson's correlation coefficient (Sokal and Rohlf 1981) to analyze the relationships among all the variables for each station. Hierarchical cluster anal-ysis, one way ANOVA, and multiple stepwise regression analysis were performed to establish the relationships between the biotic and abiotic factors and variation between months and stations using the statistical software MINITAB.

Data were transformed using the log 10 ($n + 1$) function to allow the less abun-dant species to exert same influence on the calculation of similarities (Clarke and Warwick 1994).

2.3 General Morphology of a Tintinnid ciliate

(Illustrated in Fig. 2.2)

Oral end Cavity lying at the upper end of the alimentary canal.

Aboral end The end of an animal's body opposite to its mouth.

Micronucleus The micronucleus is the smaller nucleus in ciliate protozoans which gives rise to the macronuclei and micronuclei of the individuals of the next cycle of fission.

Macronucleus Often evident, polyploid in nature, undergo direct division without mitosis.

Oral cilia Comprised of oral membranelles arranged in closed circle around the funnel-shaped oral cavity. Tintinnids use their ring of cilia around the open end both for feeding and for swimming. The cilia create currents that bring food items into the lorica and to the cell for ingestion. When the cilia are more active, they propel the tintinnids through the water.

Lorica A lorica is a shell-like protective outer covering, *often* reinforced with sand grains and osmiophilic particles collected from the surrounding water. This

Fig. 2.2 Sectional view showing the basic morphological features of a tintinnid ciliate

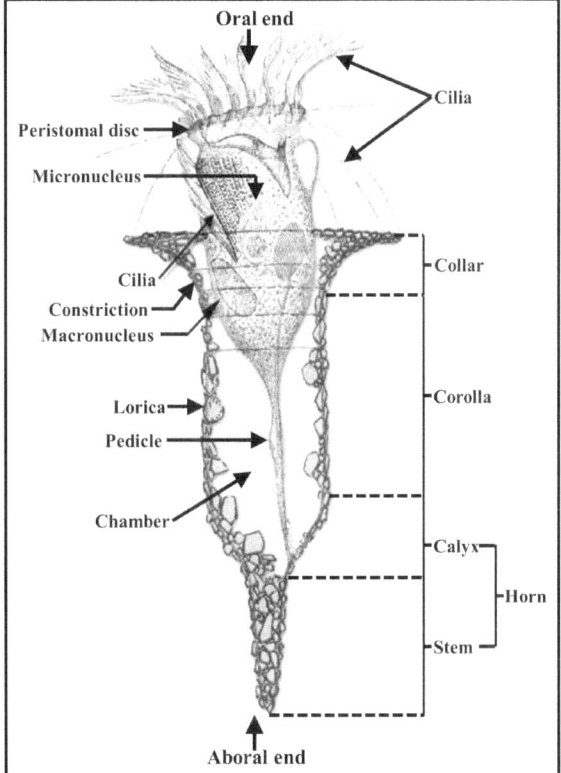

is made up of protein and usually tubular or conical in shape, with a loose case that is closed at one end. Lorica oral diameter (LOD) is a conservative taxonomic character, hence used as an important tool for taxonomic diversity. Based on lorica types, tintinnids are traditionally grouped into (i) agglutinated—composed of particles and (ii) hyaline—generally transparent loricae.

Pedicle/Peduncle A stalk-like contractile portion by which the tintinnid cell is attached at the bottom end of the lorica. During feeding, the cell extends out of the lorica and the tintinnid is propelled mouth-end forward.

2.4 Salient Features of Tintinnids Encountered in Sundarbans Coastal Regions

(A brief account of the identified tintinnids has been given in Table 2.2 along with typical morphological features)

Table 2.2 A synoptic account of the tintinnid species encountered at five sampling sites of Indian Sundarbans during 2010–2012

Family	Species	Status	Oral diameter (µm)	Length (µm)
	Agglomerated			
Codonellidae	*Tintinnopsis beroidea*	Core	12.29–38.13	31.58–118.67
	T. minuta	Core	6.25–11.84	18.29–21.54
	T. lohmani	Core	22.54–24.65	35.77–39.12
	T. cylindrica	Seasonal	65.03–71.44	288.58–311.11
	T. lobiancoi	Seasonal	10.50–11.00	20.26–25.63
	T. tocantinensis	Seasonal	50.65–52.77	165.37–190.64
	T. uruguayensis	Seasonal	20.35–21.30	42.13–44.55
	T. tubulosa	Seasonal	10.64–13.61	30.5–35.98
	T. nucula	Occasional	22.81	45.88
	T. parva	Occasional	7.25	21.89
	T. butschlii	Occasional	6.88–8.52	13.98–22.61
	T. mortensenii	Occasional	45.81	53.69
	T. kofoidi	Occasional	34–36.87	83.57–86.21
	T. directa	Seasonal	27.54–30.21	51.92–54.13
	T. brasiliensis	Occasional	23.77	39.56
	T. amoyensis	Occasional	39.58–41.67	55.87–58.36
	T. radix	Seasonal	61.98–63.77	245.39–311.86
	T. orientalis	Occasional	37.87	65.33
	T. karajacensis	Seasonal	32.67–35.88	54.18–58.44
	T. gracilis	Seasonal	39.7	9.34
	T. parvula	Occasional	41.69	18.3
	T. urnula	Occasional	22.9	5.3
	T. rotundata	Occasional	33.3	14.1
	T. acuminata	Occasional	41.1	22.65

(continued)

Table 2.2 (continued)

Family	Species	Status	Oral diameter (µm)	Length (µm)
	Agglomerated			
Tintinnidiidae	*Leprotintinnus simplex*	Seasonal	18.72–46.87	57.59–184.79
	L. nordqvisti	Seasonal	24.66–30.41	54.89–59.21
	Tintinnidium primitivum	Seasonal	8.09–12.67	25.82–37.39
	T. mucicola	Seasonal	9.82–11.50	30.58–37.48
Codonellopsidae	*Codonellopsis schabi*	Occasional	34.88–38.24	67.41–70.32
	Stenosemella ventricosa	Seasonal	23.75	24.14
Dictyocystidae	*Wangiella dicollaria*	Occasional	27.75	14.43
	Non-agglomerated			
Xystonellidae	*Favella ehrenbergii*	Occasional	92.45–95.72	292.76–322.5
Tintinnidae	*Amphorellopsis acuta*	Seasonal	31.68–37.94	90.98–93.16
	Eutintinnus sp.	Seasonal	13.63	34.25
Metacylididae	*Metacylis mediterranea*	Occasional	28.3	37.25
	Metacylis sp.	Occasional	38.41–41.60	68.11–70.75
	Helicostomella sp.	Occasional	41.55	16.4

2.4.1 Tintinnopsis beroidea (Stein, 1867)

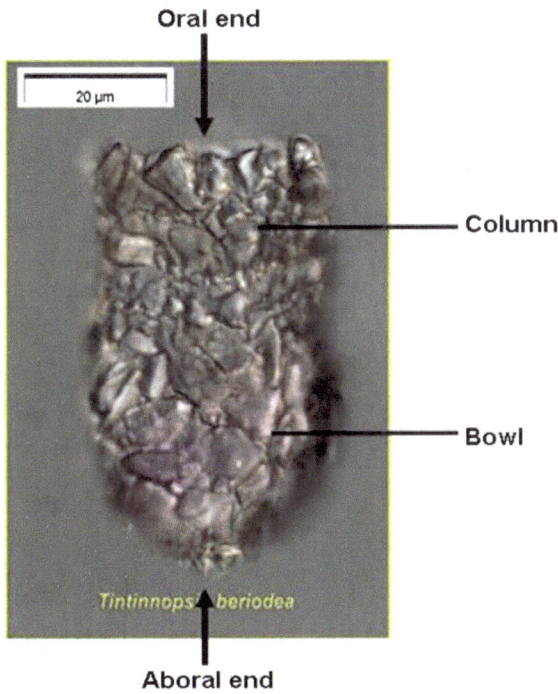

Phylum: Ciliophora
Subphylum: Intramacronucleata
Class: Spirotrichea
Subclass: Choreotrichia
Order: Tintinnida
Family: Codonellidae

Morphological characteristics Lorica bullet-shaped, usually cylindrical in the anterior portion, aborally conical (75°–85°), oral rim ragged; aboral end acute or bluntly pointed; wall rather coarse, 0.03–0.04 oral diameters in thickness, without spiral structure.

Measurements The length varied throughout the year, depending on the season. The smallest loricae were observed in winter ranging from 13 to 17 µm when the water temperature was ~21 °C. In contrast, during autumn and summer, the length of the lorica ranged between 75 and 118 µm with temperatures ~31 °C. Oral diameters were variable, ranging between 12.29 and 30.13 µm.

Distribution This species is cosmopolitan in nature, recorded mostly from Gulf of Mexico (Federal and Camp 2010), Newport (Town) (De Pauw 1975), Plymouth (Hayward and Ryland 1990). *Tintinnopsis beroidea* was registered in relatively low abundance in the northwestern waters of the Arabian Gulf. The species is considered the most common and dominant species in different coastal region of India.

Comments Despite the fact that this species appeared within a wide range of salinities (between 18 and 35 p.s.u.), it was often more abundant at salinities >25 p.s.u. (Urrutxurtu 2004). This species is known to produce cysts that sink and deposit in the sediments until suitable environmental conditions trigger the return to a viable form. Since the light conditions have been found to play a critical role in the process, it is reasonable that the summer insolation will have induced a progressive excystment phase of *T. beroidea* on the shallow seabed (Sitran et al. 2009).

2.4.2 Tintinnopsis minuta (Wailes, 1925)

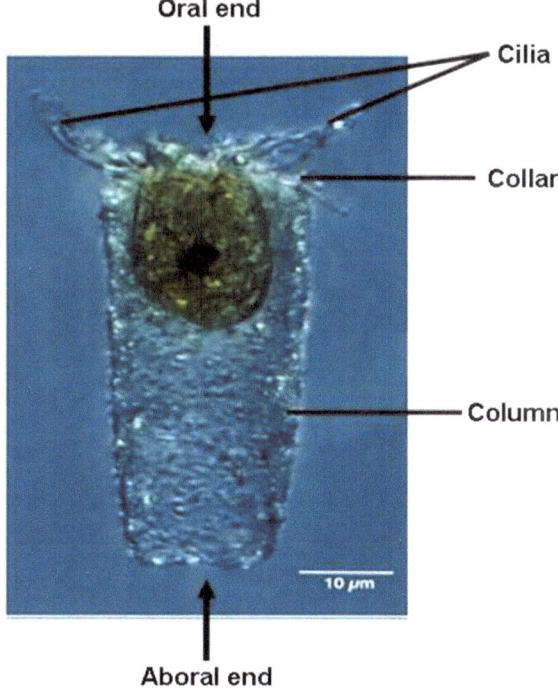

Phylum: Ciliophora
Subphylum: Intramacronucleata
Class: Spirotrichea
Subclass: Choreotrichia
Order: Tintinnida
Family: Codonellidae

Morphological characteristics Lorica was not divided into collar and bowl. It is minute, conical shaped with a blunt end. Agglomeration was coarse and even.

Measurements Smallest tintinnids present in coastal regions of Sundarbans with a length ranging from 18.29 to 21.54 μm and an oral diameter ranging from 6.25 to 11.84 μm.

Distribution Recorded from Wimereux (Muller 2004), Gulf of Mexico (Federal and Camp 2010), Central Long Island Sound (Capriulo and Carpenter 1983) and Southeast coast of India (Godhantaraman 2002).

Comments This species was present almost throughout the year reaching in highest densities in May 2010 in Dhamakhali of Indian Sundarbans and April 2010 in Canning (250 ind l^{-1}). Particularly in summer months, it accounted for 50–60 % of total tintinnid abundance (Godhantaraman 2001). The highest density was found when the water temperature was between 28 and 30 °C. It is generally a high temperature and wide salinity species (Godhantaraman 2001) and thus was observed in a broad range of salinity (6.3–26.6 p.s.u).

2.4.3 *Tintinnopsis lohmani (Laackmann, 1906)*

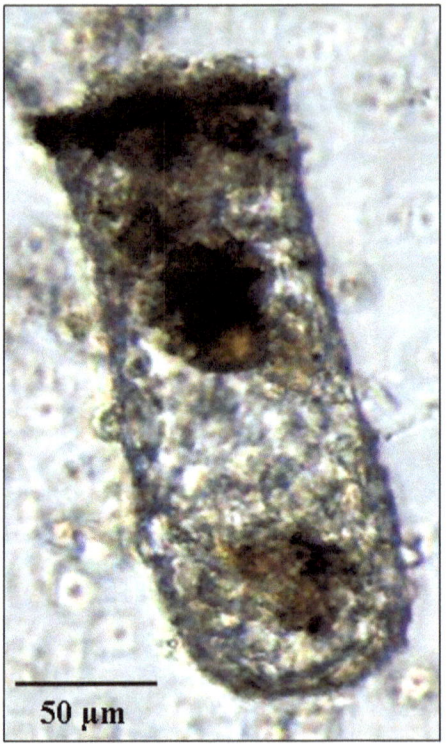

50 µm

Phylum: Ciliophora
Subphylum: Intramacronucleata
Class: Spirotrichea
Subclass: Choreotrichia
Order: Tintinnida
Family: Codonellidae

Morphological characteristics Lorica vase-like with a cylindrical collar; collar usually short, 0.16–0.40 µm of the total length; bowl expanding; aboral region rounded or convex conical; wall coarsely agglomerated, a few spiral turns appearing in the collar.

Measurements Length 35.77–75.2 µm; oral diameter 12.9–24.65 µm.

Distribution Abundant in the cool season in the waters surrounding Bubiyan Island, Kuwait (Yamani et al. 2011), rather it is more common in Indian coastal waters (Zipcodezoo.com).

Comments It is a seasonal species which was observed from early premonsoon and monsoon period (June–September), reaching in a highest concentration of 166 ind l^{-1} in August 2012 at Gangasagar of Indian Sundarbans. Maximum abundance coincided with temperature ~28 °C and salinity within ~26 p.s.u.

2.4.4 *Tintinnopsis cylindrica (Daday, 1887)*

Tintinnopsis cylindrica

50 μm

Phylum: Ciliophora
Subphylum: Intramacronucleata
Class: Spirotrichea
Subclass: Choreotrichia
Order: Tintinnida
Family: Codonellidae

Morphological characteristics Lorica cylindrical, oral margin usually smooth without any flare. Aboral portion tapering ending with a stout aboral horn. The wall was very coarsely agglomerated.

Measurements Lorica length varied from 288.58 to 311.11 μm. Oral diameter remained relatively invariable, ranging from 65 to 70 μm. The pedicel was hollow and normally appeared broken. Its length ranged between 15 and 40 mm, but normally it was ~25 μm long.

Distribution Recorded from Lake Nakaumi, China (Uye and Godhantaraman 2003), Cochin backwaters, south India (Jyothibabu et al. 2006) and Vellar estuary, south India (Godhanraman 2001).

Comments Generally observed in peak summer and early autumn, at temperatures between 21.7 and 32.8 °C. The highest abundance, ~75 ind l^{-1}, was registered at Chemaguri (S_4) of Indian Sundarbans, when salinity values were >21 p.s.u.

2.4.5 *Tintinnopsis lobiancoi (Daday, 1887)*

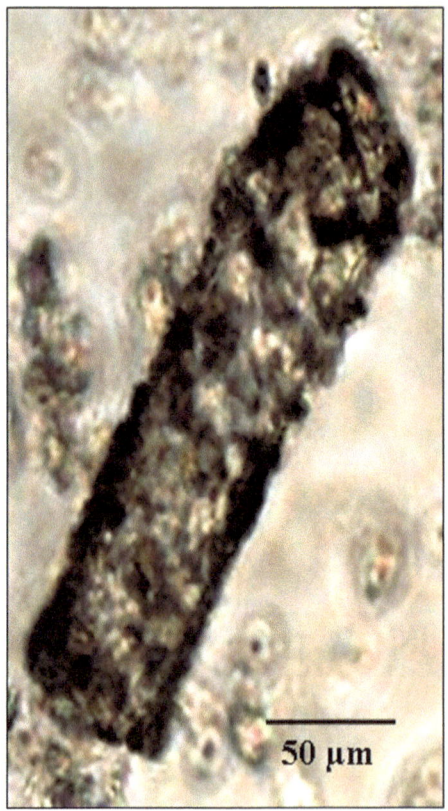

Phylum: Ciliophora
Subphylum: Intramacronucleata
Class: Spirotrichea
Subclass: Choreotrichia
Order: Tintinnida
Family: Codonellidae

Morphological characteristics Lorica elongate, tubular, usually straight, oral rim ragged; aboral end rounded or shaped somewhat irregularly; wall agglomerated roughly, but comparatively thin without a spiral structure.

Measurements Length of the lorica ranged from 20.26 to 25.63 µm and the oral diameter ranged from 10.50 to 11.0 µm

Distribution Predominant species in Asia recorded from Lake Nakaumi, Japan (Uye and Godhantaraman 2003), Northern China (Jiang et al. 2012) and Southeast coast of India (Godhantaraman 2002).

Comments It is a seasonal species which was observed from late monsoon to early postmonsoon period (August–October), reaching in a highest concentration of 100 ind l^{-1} in October 2010 at Lot 8 (S_3) of Indian Sundarbans. The highest numerical density coincided with temperature between 31 and 33 °C and salinity within 15 p.s.u.

2.4.6 *Tintinnopsis tocantinensis (Kofoid and Campbell, 1929)*

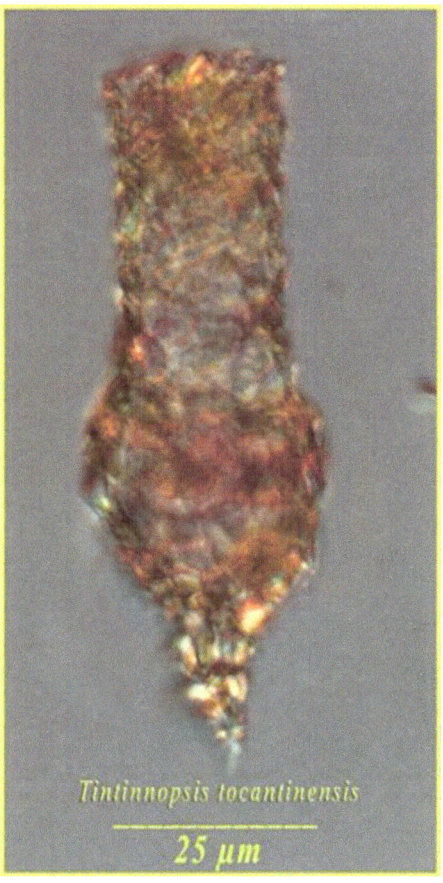

Phylum: Ciliophora
Subphylum: Intramacronucleata
Class: Spirotrichea
Subclass: Choreotrichia

Order: Tintinnida
Family: Codonellidae

Morphological characteristics Lorica elongated, cylindrical anteriorly, expanding posteriorly, tapering distally into a stout aboral horn; dilated part not spiraled, aboral horn conical, obliquely or irregularly open at the tip.

Measurements Total length 165.37–190.64 µm. Oral diameter 50.65–52.77 µm.

Distribution Common in the northwestern waters of the Arabian Gulf (Yamani et al. 2011), Gulf of Mexico (Felder and Camp 2010), coastal waters of China (Jiang et al. 2012) and India (Naha Biswas et al. 2013).

Comments Seasonal species, generally found from late monsoon to early post-monsoon period. Maximum abundance (50 ind l^{-1}) was found in Canning of Indian Sundarbans during August 2010 coincided with the water temperature of 30.7 °C and salinity 19.98 p.s.u.

2.4.7 *Tintinnopsis uruguyansis (Nie and Cheng, 1947)*

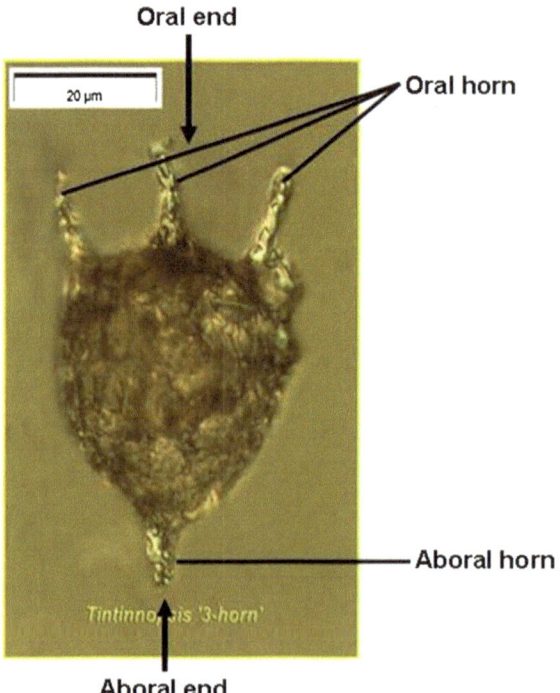

Phylum: Ciliophora
Subphylum: Intramacronucleata
Class: Spirotrichea
Subclass: Choreotrichia
Order: Tintinnida
Family: Codonellidae

Morphological characteristics Lorica is divided into two parts—A very short column and a spherical bowl. Short aboral horn was present at the end of the lorica. Mouth flaring with two to three "horn"-like structures for which it was also denoted as *Tintinnopsis* 3 horn.

Measurement The length of the lorica did not vary much, ranging between 42.13 and 44.55 μm. The oral diameter ranged from 20.35 to 21.30 μm. The aboral horn was much short ranging between 3 and 4 μm. Approximate ratio between length and oral diameter was ~1.0:2.0.

Distribution Recorded from Central Long Island Sound (Capriulo and Carpenter 1983), Cochin backwater, India (Jyothibabu et al. 2006) and Vellar estuary, India (Godhantaraman 2001).

Comments It is a postmonsoon species, found only during October–November of each year. The species seems to prefer the temperature between 28 and 30 °C and salinity below 15 p.s.u. The shape and size of this species varied slightly between different seasons.

2.4.8 *Tintinnopsis tubulosa (Levander, 1900)*

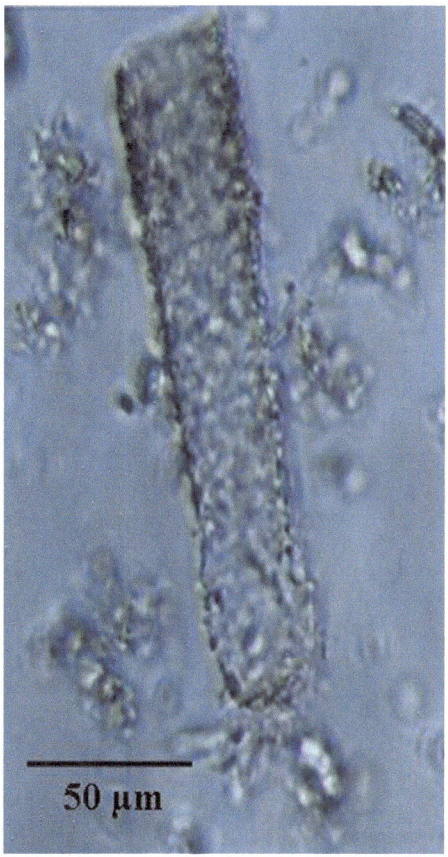

Phylum: Ciliophora
Subphylum: Intramacronucleata
Class: Spirotrichea
Subclass: Choreotrichia
Order: Tintinnida
Family: Codonellidae

Morphological characteristics Lorica is differentiated into two regions: a cylin-drical collar and a bowl, its length 2.1–3.5 oral diameters; collar 0.25–0.50 of the total length long; bowl somewhat inflated, broadest in the posterior 0.33–0.40 of the lorica, its greatest transdiameter 1.05–1.25 oral diameters; aboral region usu-ally conical (75°–90°) to an acute distal end or rarely rounded with a blunt end; wall rather thin, but irregular in appearance, no spiral structure.

Lorica is differentiated into two regions—a long collar and a short bowl at the end. Oral flare absent. Bowl and column uniformly agglomerated.

Measurement The length of the lorica varied from 30.5 to 35.98 μm and the oral diameter ranged from 10.64 to 13.61 μm.

Distribution Recorded from Gulf of Mexico (Federal and Camp 2010), Japan, Portugal and Spain (Zipcodezoo.com). This species is highly common in Indian coastal waters found from Southeast coast of India (Godhantaraman 2002), Cochin backwaters (Jyothibabu et al. 2008), Bahuda estuary, East coast of India (Mishra and Panigrahy 1999), Porto novo region, India (Krishnamurthy and Santhanam 1975).

Comments The species was recorded mostly during postmonsoon followed by premonsoon. The maximum abundance (100 ind l^{-1}) was noticed during February 2010 at Dhamakhali (S$_{12}$) of Indian Sundarbans when the temperature was 25 °C and the salinity was 14.51 p.s.u.

2.4.9 Tintinnopsis nucula (Fol, 1884)

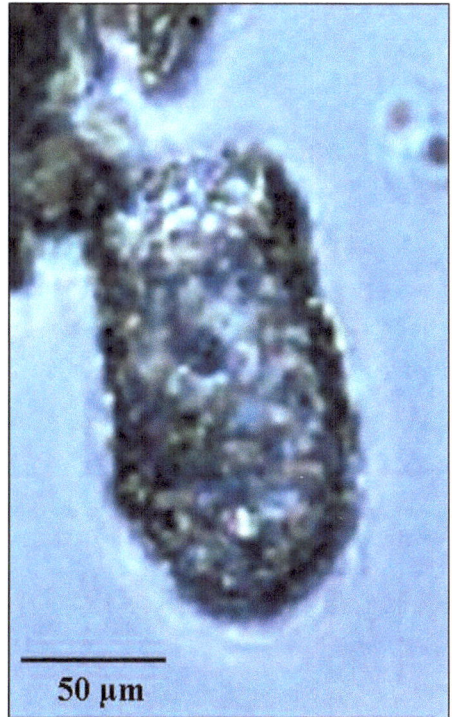

Phylum: Ciliophora
Subphylum: Intramacronucleata
Class: Spirotrichea
Subclass: Choreotrichia
Order: Tintinnida
Family: Codonellidae

Morphological characteristics Lorica oval shaped with rounded aboral end. Agglomeration of the lorica was even and coarse. Mouth without oral flare.

Measurement Lorica length was found to be ~45.8 μm, and oral diameter was around 23 μm.

Distribution Recorded from Wimereux, Belgium (Muller 2004), Southeast coast of India (Godhantaraman 2002).

Comments Occasional species found only 2–3 times throughout the study period. Maximum abundance (50 ind l^{-1}) was recorded during November 2010 at Canning (S$_3$) of Indian Sundarbans when the temperature and salinity were 26.75 °C and 14.9 p.s.u., respectively.

2.4.10 Tintinnopsis parva (Merkle, 1909)

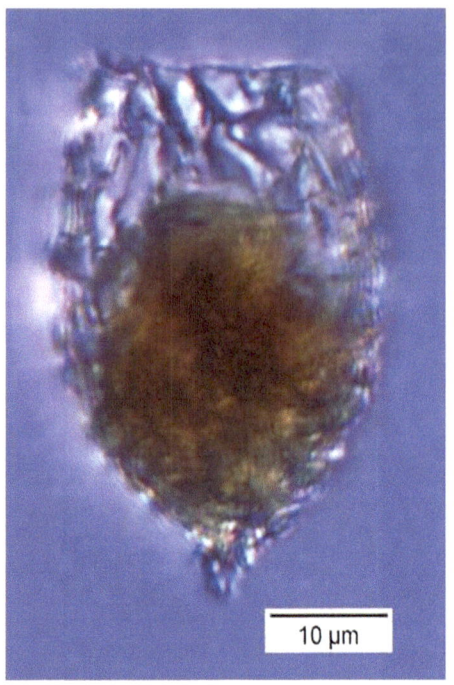

10 μm

Phylum: Ciliophora
Subphylum: Intramacronucleata
Class: Spirotrichea
Subclass: Choreotrichia
Order: Tintinnida
Family: Codonellidae

Morphological characteristics Lorica small, widest in middle, narrowing slightly to mouth, more sharply to pointed aboral end.

Measurements It is a comparatively smaller species with the length of 21.89 μm, and the oral diameter was 7.25 μm.

Distribution Recorded from Wimereux (Muller 2004), northwestern waters of the Arabian Gulf (Yamani et al. 2011) and Southeast coast of India (Godhantaraman 2002).

Comment Found in comparatively cool season during November–January 2011 mainly in Canning of Indian Sundarbans. Highest abundance reached up to 50 ind l^{-1} when the water temperature was ~19.01 °C and the salinity was 15.3 p.s.u.

2.4.11 *Tintinnopsis butschlii (Daday, 1887)*

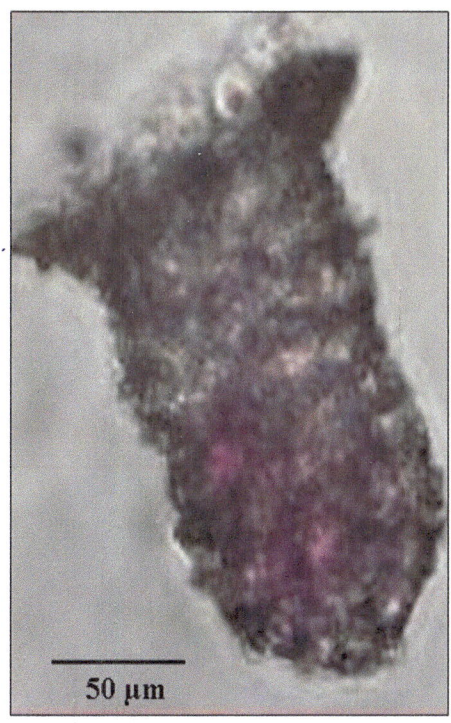

50 μm

Phylum: Ciliophora
Subphylum: Intramacronucleata
Class: Spirotrichea
Subclass: Choreotrichia
Order: Tintinnida
Family: Codonellidae

Morphological characteristics Lorica cylindrical in shape with a medium-sized oral flare. Aboral end sharply pointed without any horn. Agglomeration was coarse and even.

Measurement Lorica length ranged from 13.98 to 22.61 μm, whereas the LOD did not vary much, ranging from 6.88 to 8.5 μm.

Distribution Recorded from Indian coastal waters.

Comments Occasional species found to prefer the premonsoon season with temperature ~30–32 °C and salinity >20 p.s.u. Maximum abundance (100 ind l^{-1}) was recorded at Lot 8 of Indian Sundarbans during March 2010.

2.4.12 *Tintinnopsis mortensenii (Schmidt, 1901)*

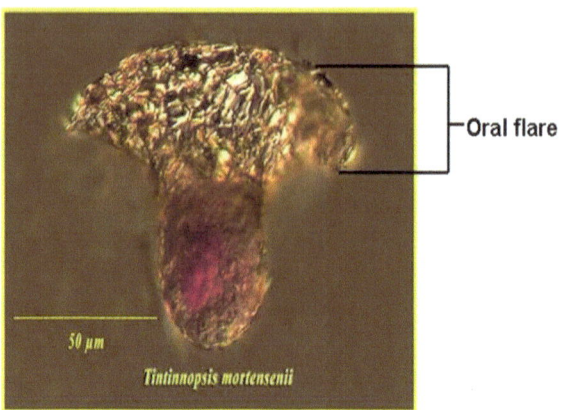

Phylum: Ciliophora
SubphylumL: Intramacronucleata
Class: Spirotrichea
Subclass: Choreotrichia
Order: Tintinnida
Family: Codonellidae

Morphological characteristics Lorica bell-shaped with slightly concave collar, very wide, horizontal oral funnel, and rounded aboral end.

Measurements Length 45.81 µm; oral diameter 53.69 µm;

Distribution Commonly recorded from the Gulf of Mexico (Felder and Camp 2010), northwestern waters of the Arabian Gulf (Yamani et al. 2011), Southeast coast of India (Godhantaraman 2002), Cochin backwater (Jyothibabu et al. 2006; Mishra and Panigrahy 2006).

Comments Absolutely occasional in Sundarbans coastal region, found only twice throughout the study period. The maximum abundance reached to 75 ind l^{-1} in March 2010 at Lot 8 of Indian Sundarbans which coincided with ~32.85 °C water temperature and ~21.3 p.s.u salinity.

2.4.13 *Tintinnopsis kofoidi (Hada, 1932)*

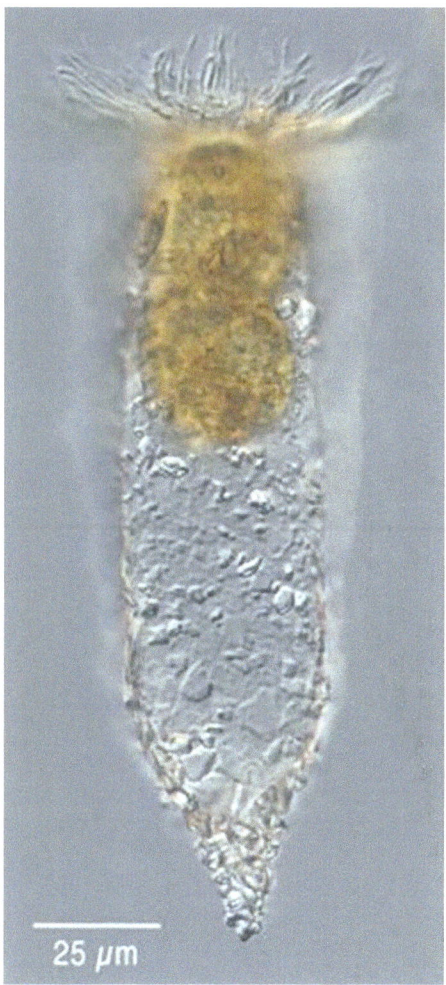

Phylum: Ciliophora
Subphylum: Intramacronucleata
Class: Spirotrichea
Subclass: Choreotrichia
Order: Tintinnida
Family: Codonellidae

Morphological characteristics Lorica bullet-shaped, usually cylindrical in the anterior 0.6–0.7 of the total length, aboral end opened. Oral rim ragged; wall rather coarse, without spiral structure. A small horn present at the end of the lorica.

Measurement Length of the lorica varied from 83.57 to 86.21 µm and the oral diameter ranged from 34 to 36.87 µm.

Distribution Recorded from Florida (Cosper 1972), Gulf of Mexico (Felder and Camp 2010), South Atlantic (Balech 1951), Pichavaram mangrove, India (Godhantaraman 1994).

Comments This species was found to exist from late postmonsoon (February) to early premonsoon (around mid-April) period. But it was an occasional species in this specific mangrove wetland, and the maximum abundance (75 ind l^{-1}) was recorded during March 2010 at Lot 8 of Indian Sundarbans when the temperature was ~32.85 °C and the salinity was 21.3 p.s.u.

2.4.14 *Tintinnopsis directa (Hada, 1932)*

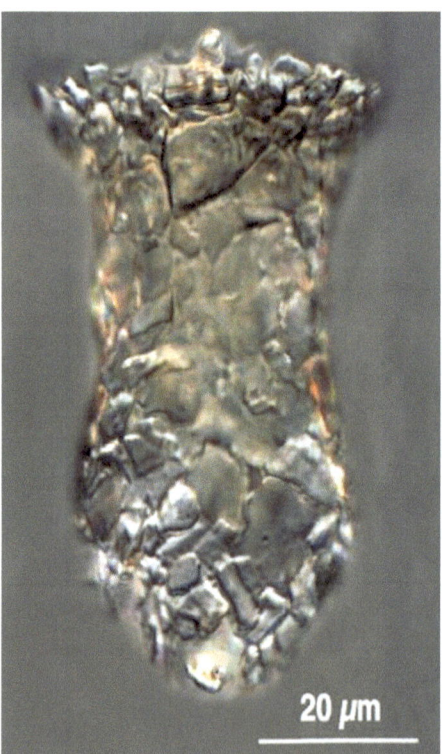

Phylum: Ciliophora
Subphylum: Intramacronucleata
Class: Spirotrichea
Subclass: Choreotrichia
Order: Tintinnida
Family: Codonellidae

Morphological characteristics The species is characterized by the presence of a moderately erect lorica which is campanulate anteriorly and subspherical posteriorly. The lorica is about 1.7 oral diameters in length. Agglomeration is light on the cylindrical part of the lorica and fairly pronounced on the bowl. Oral rim irregular, flaring suboral region somewhat tapering, conical, laid up with about 6 spiral turns, narrowest at the basal portion of the subcylindrical part, wall rather coarse in the posterior part at the thickest portion of the aboral region, aboral end hemispherical; suboral region somewhat tapering, conical (5°–10°), laid up with about 6 spiral turns, narrowest at the basal portion of the subcylindrical part, its smallest transdiameter 0.68–0.82 of the oral diameter posterior region subspherical, with a rounded aboral end, 0.80–0.95 oral diameters in transdiameter; wall rather coarse in the posterior part, about 0.035 diameters in thickness at the thickest portion of the aboral region.

Measurements Total length and the oral diameter varied from 27.54 to 30.21 μm and from 51.92 to 54.13 μm, respectively.

Distribution Found to be common in Gulf of Mexico (Federal and Camp 2010), in the northwestern waters of the Arabian Gulf, in the central region of the Gulf, mainly along the Iranian coast (Al Yamani et al. 2011), Southeast coast of India (Godhantaraman 2002), Cochin backwaters (Jyothibabu et al. 2008) and Bahuda estuary (Mishra and Panigrahy 1999).

Comments It is a premonsoon species recorded maximum during March 2010 when the temperature was 32.85 °C and the salinity was 21.3 p.s.u.

2.4.15 *Tintinnopsis brasiliensis (Kofoid and Campbell, 1929)*

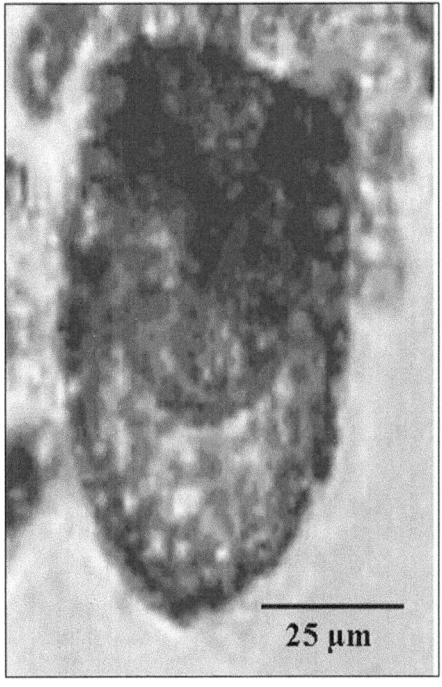

25 μm

Phylum: Ciliophora
Subphylum: Intramacronucleata
Class: Spirotrichea
Subclass: Choreotrichia
Order: Tintinnida
Family: Codonellidae

Morphological characteristics Body generally conical in shape, truncate at the anterior end, the buccal area is wide. *Tintinnopsis directa* (Hada, 1932) posterior portion narrowed and always forming a short stalk, with which the cell adheres to the inside of the lorica. Buccal cavity prominent and deep. Cilia of membranelles. Shape rather constant, subconical with posterior portion blunt to slightly tapering, without collar margin of lorica opening with some particles forming an uneven edge. Wall thin and opaque, with coarse surface due to outer covering, which is composed of many large, irregularly shaped mineral-like particles.

Measurements Length 23.77 μm and oral diameter 39.56 μm.

Distribution Recorded from Taiping Cape of Qingdao, China (ShengfangTsai et al. 2006).

Comments *T. brasiliensis* was registered during premonsoon (April 2010) at Lot 8 of Indian Sundarbans, when the temperature and salinity were 31 °C and 20.1 p.s.u, respectively.

2.4.16 *Tintinnopsis amoyensis (Nie, 1934)*

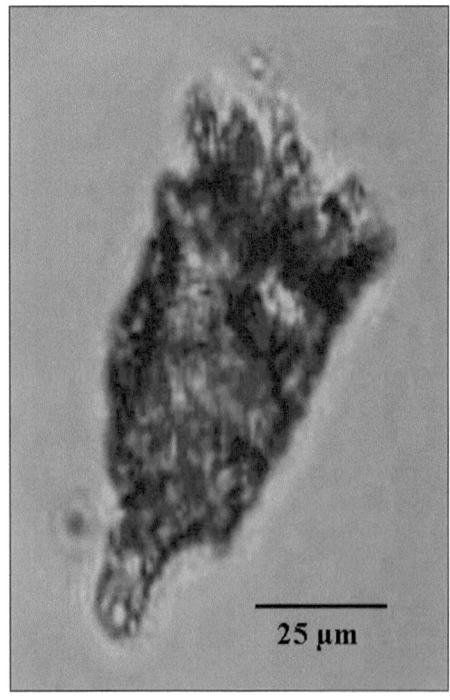

25 μm

Phylum: Ciliophora
Subphylum: Intramacronucleata
Class: Spirotrichea
Subclass: Choreotrichia
Order: Tintinnida
Family: Codonellidae

Morphological characteristics Lorica thistle-funnel shaped, oral rim very roughened; collar flaring, inverted conical; bowl subcylindrical anteriorly, convex conical at aboral region; aboral horn stout, short, obliquely or irregularly opened at the tip; wall thick and coarsely agglomerated.

Measurements Length of the lorica varied from 55.87 to 58.36 μm and the oral diameter ranged from 39.58 to 41.67 μm.

Distribution Scarce in Indian coastal waters.

Comments *T. amoyensis* is a small species, and it seems to be more closely related to *T. meunieri* Kofoid and Campbell; however, the lorica of the present species is more slender in form and much smaller in size. Species was registered during March 2010 at the site Lot 8 of Indian Sundarbans.

2.4.17 Tintinnopsis radix (Imhof, 1886)

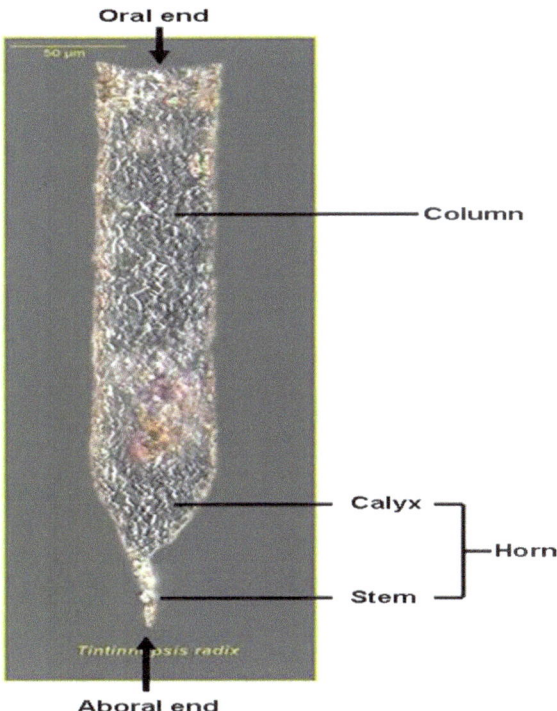

Phylum: Ciliophora
Subphylum: Intramacronucleata
Class: Spirotrichea
Subclass: Choreotrichia
Order: Tintinnida
Family: Codonellidae

Morphological characteristics Lorica elongate, slender, tubular, oral rim generally entire (smooth and round) or sometimes irregular; bowl long, cylindrical; aboral region tapering gradually into an aboral horn, usually more or less curved, with an irregularly formed aboral opening typically set laterally as gouged, wall thin and fragile, with a variable spiral structure.

Reported measurements The range of total length varied from 245.39 to 311.86 µm. Oral diameter ranged from 61.98 to 63.77 µm.

Distribution *Tintinnopsis radix* is common and abundant at Gulf of Mexico (Federal and Camp 2010), Rijeca Bay, Croatia (Zavodnik and Kovacic 2000), all the studied locations in the Arabian Gulf, from Bubiyan Island in the north to the Strait of Hormuz in the south (Yamani et al. 2011) and all studied Indian Coastal waters from Porto Novo region east coast of India (Krishnamurthy and Santhanam 1975) to Parangipettai Southeast coast of India (Godhantaraman 2002).

Comments It is completely a postmonsoon species which was noticed for the first time in Gangasagar of Indian Sundarbans during November 2010 coinciding with a sudden outburst of phytoplankton population dominated by a centric diatom *Hemidiscus hardmannianus*. The highest abundance of this species was found to be 200 ind l^{-1} during February 2010 which coincided with 23.4 °C water temperature and 21.8 p.s.u salinity.

2.4.18 *Tintinnopsis orientalis (Kofoid and Campbell, 1929)*

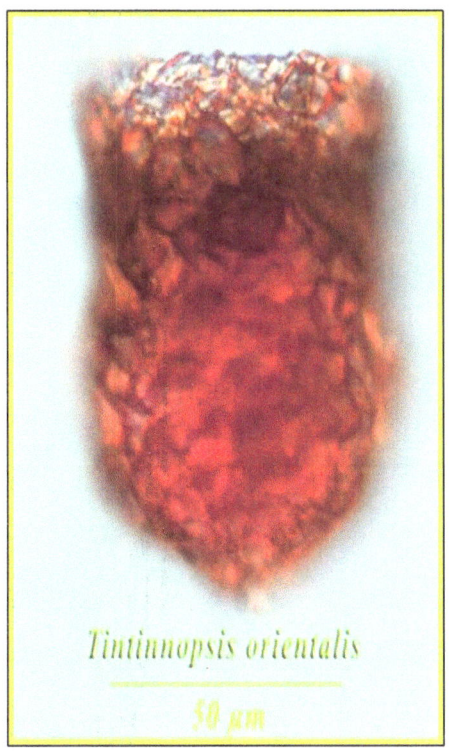

Phylum: Ciliophora
Subphylum: Intramacronucleata
Class:Spirotrichea
Subclass: Choreotrichia
Order: Tintinnida
Family: Codonellidae

Morphological characteristics Lorica with distinct collar; collar 1/4–1/3 length of the bowl, convex outwardly; bowl acorn shaped, widest near or shortly below its middle; aboral end baggy, not flattened, with or without a faintly emergent point.

Reported measurements Length 65.33 μm; oral diameter 37.87 μm.

Distribution *Tintinnopsis orientalis* is relatively common in the northwestern waters of the Arabian Gulf during the warm season (Yamani et al. 2011) and Northern China (Jiang et al. 2011).

Comments Occasional species found only once during March 2010 in Lot 8 of Indian Sundarbans with an abundance of 75 ind l^{-1}.

2.4.19 *Tintinnopsis karajacensis (Brandt, 1896)*

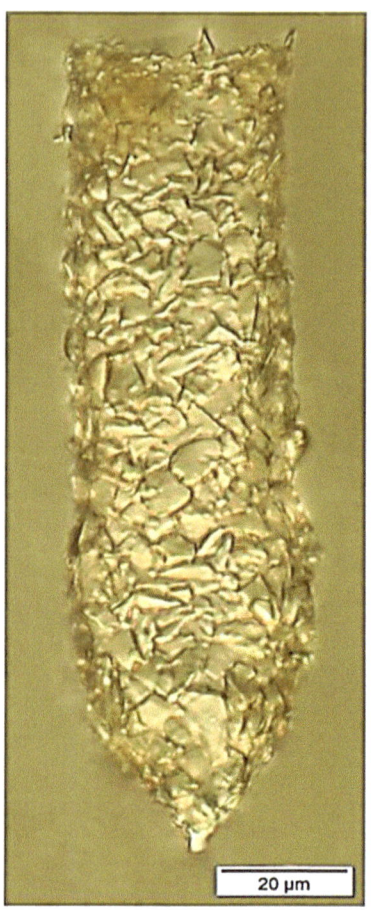

Phylum: Ciliophora
Subphylum: Intramacronucleata
Class: Spirotrichea
Subclass: Choreotrichia
Order: Tintinnida
Family: Codonellidae

Morphological characteristics Lorica capsular; oral margin roughened; bowl cylindrical in the margin part, rounded in the aboral end; wall coarsely agglomerated, often with a spiral structure in the anterior region of the lorica; aboral end rounded or disfigured as the result of irregularly agglomerated particles; wall coarse, having several slight spiral turns in the anterior half.

Reported measurements Length 60–172 μm; oral diameter 30–64 μm; approximate ratio L/oral diameter 2.0–2.7

Distribution *Tintinnopsis karajacensis* was registered in relatively low abundance in the northwestern Arabian Gulf (Yamani et al. 2011), Lake Nakaumi, Japan (Uye and Godhantaraman 2003), Central Long Island Sound (Capriulo and Carpenter 1983), Vellar estuary, India (Godhantaraman 2001) and Nervion estuary, Spain (Urrutxurtu 2004).

Comments This is an occasional species which differs from *T. directa* in the absence of the oral flare and the posterior inflation, from *T. cochleata* in less extensive spiral organization and in roughened agglomeration, from *T. lobiancoi* in the shorter lorica, and from *Tintinnopsis rotundata* in more slender proportions and in the shape of the aboral end.

2.4.20   Tintinnopsis gracilis *(Kofoid and Campbell, 1929)*

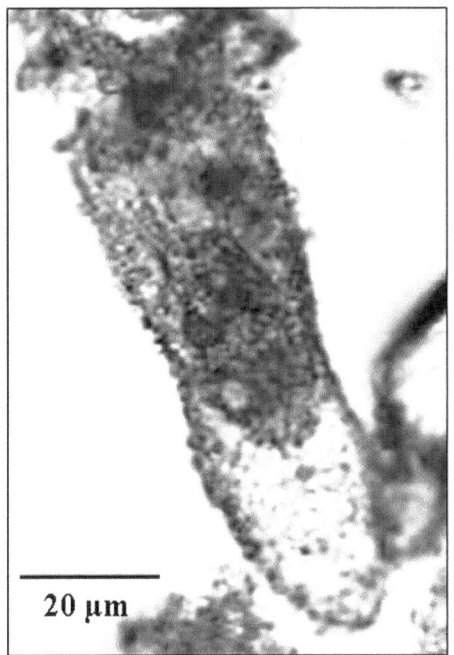

Phylum: Ciliophora
Subphylum: Intramacronucleata
Class: Spirotrichea
Subclass: Choreotrichia
Order: Tintinnida
Family: Codonellidae

Morphological characteristics Lorica finger-shaped, oral margin usually comparatively smooth; bowl tubular, sometimes slightly swollen in the posterior one-third of the bowl; aboral region convex conical with a blunt distal end; wall coarsely agglomerated without a spiral structure.

Measurements Length 105–125 μm; oral diameter 28–34 μm; approximate ratio of length and oral diameter was found to be 3.3–4.0.

Distribution *Tintinnopsis gracilis* is one of the most common and abundant species in the Northwestern Arabian Gulf and in the central region of the Gulf, mainly along the Iranian coast (Al Yamani et al. 2011), Gulf of Mexico (Federal and Camp 2010), Asia (Japan) [Zipcodezoo.com] and Indian coastal waters.

Comments The species differs from *T. karajacensis* Brandt in the conical aboral region instead of the round.

2.4.21 *Tintinnopsis parvula (Jörgensen, 1912)*

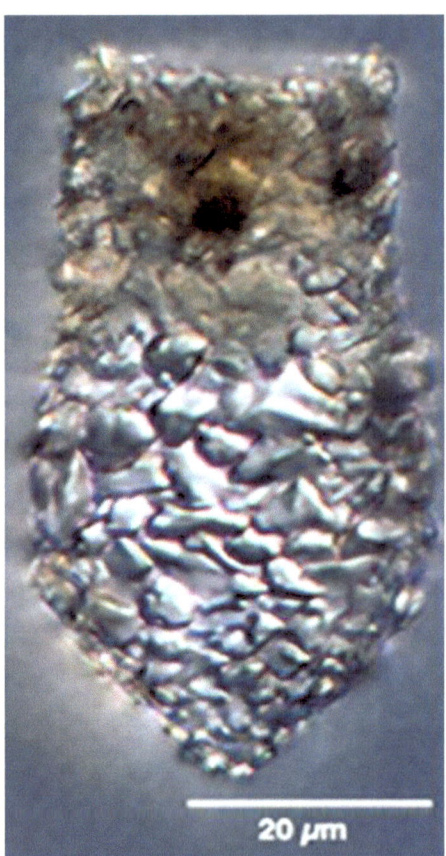

Phylum: Ciliophora
Subphylum: Intramacronucleata
Class: Spirotrichea
Subclass: Choreotrichia
Order: Tintinnida
Family: Codonellidae

Morphological characteristics Lorica small, slightly expanded below a cylindrical anterior region, pointed aborally.

Measurements Length 51–70 μm; oral diameter 18.3–29 μm.

Distribution Common and abundant in the northwestern waters of the Arabian Gulf and in the waters surrounding Bubiyan Island.

Comments Lorica more pointed aborally than *T. nucula* and more expanded than *T. beroidea*. *T. parvula* is an occasional species, found during early monsoon period with a maximum abundance of 50 ind l^{-1} at the site Gangasagar (S$_7$) of Indian Sundarbans.

2.4.22 *Tintinnopsis urnula (Meunier, 1910)*

Phylum: Ciliophora
Subphylum: Intramacronucleata
Class: Spirotrichea
Subclass: Choreotrichia
Order: Tintinnida
Family: Codonellidae

Morphological characteristics Lorica campanulate; oral rim ragged; bowl constricted at the suboral 1/3 of the total length, sides concave in the anterior 2/3 of the lorica; aboral end subacute; wall more or less coarse, with very slight spiral organization in the suboral part.

Measurement Length 58.2 μm; oral diameter 41.1 μm.

Distribution The species is very much abundant in East China Sea.

Comments Differs from *T. beroidea* Stein in the presence of the suboral constriction. *Tintinnopsis urnula* is very rare in HRE water and documented only at Lot no. 8 of Indian Sundarbans with an abundance of 83 ind l^{-1} during premonsoon (June 2012) coinciding by moderate temperature (30 °C) and salinity (11.5 p.s.u).

2.4.23 *Tintinnopsis rotundata (Kofoid and Campbell, 1929)*

Phylum: Ciliophora
Subphylum: Intramacronucleata
Class: Spirotrichea

Subclass: Choreotrichia
Order: Tintinnida
Family: Codonellidae

Morphological characteristics   Lorica cylindrical with hemispherical aboral end; oral margin usually ragged; wall comparatively thin, thickly encrusted with particles, spiral structure invisible.

Measurement   Length 62 μm; oral diameter 38 μm;

Distribution   *Tintinnopsis rotundata* was registered in low abundance in the northwestern waters of the Arabian Gulf.

Comments   This species differs from the typical form of *T. karajacensis* Brandt in the possession of the thin wall without spiral structure. *Tintinnopsis rotundata* was registered in low abundance (41 ind 1^{-1}) in the high-energy zone Gangasagar of Indian Sundarbans during January 2013 assessed by low temperature (19 °C) and high salinity (22 p.s.u).

2.4.24   Tintinnopsis acuminata (Daday, 1887)

10 μm

Phylum: Ciliophora
Subphylum: Intramacronucleata
Class: Spirotrichea
Subclass: Choreotrichia
Order: Tintinnida
Family: Codonellidae

Morphological characteristics Lorica tubular, oral rim ragged, aboral region in lower 1/4 conical, aboral end blunt. Wall without spiral structure, with spare agglomeration. Oral region has no oral funnel.

Measurement Length 63.75 μm; oral diameter 22.65 μm.

Distribution *Tintinnopsis acuminata* was registered in relatively low abundance in the northwestern Arabian Gulf.

Comments: *Tintinnopsis acuminata* was registered only in Gangasagar of Indian Sundarbans with a maximum abundance of 12 ind l^{-1} during monsoon period (October 2012) coincided by temperature <30 °C and salinity <20 p.s.u.

2.4.25 *Leprotintinnus simplex (Schmidt, 1901)*

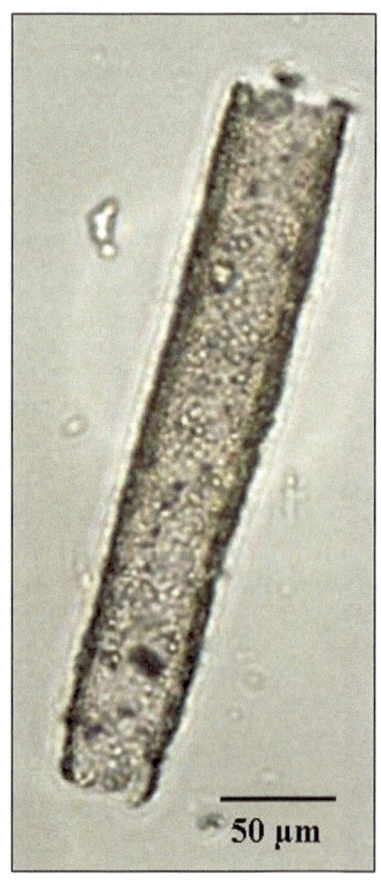

50 μm

Phylum: Ciliophora
Subphylum: Intramacronucleata
Class: Spirotrichea
Subclass: Choreotrichia
Order: Tintinnida
Family: Tintinnidiidae

Morphological characteristics Lorica soft and fragile, cylindrical, never with flaring oral or aboral ends sometimes slightly narrowing toward the aboral end. Wall often shows a weak spiral structure, more pronounced at aboral half of lorica, and it usually depends both on the shape of the wall and of the distribution of particles. Agglomeration is relatively poor and even at the oral part of lorica and denser at the aboral part.

Measurement Total body length ranges from 57.59 to 184.79 μm, and the oral diameter ranges from 18.72 to 46.87 μm. Lorica is very fragile, especially in the aboral part, such that the minimum size of this tintinnid is not identifiable. Usually, there are many fragments of loricas in the samples (loricas could be broken during sampling or transportation), which makes it difficult to determine the real size of tintinnids.

Distribution Recorded mostly from India, especially from Parangipettai, Southeast coast of India (Godhantaraman 2002) and Cochin backwaters, Southwest coast of India (Jyothibabu et al. 2008). It was also found to be abundant in northern waters of Kuwait of Boubiyan Island but only in winter and early spring. In spring, this tintinnid produces one or two cysts and survives the warm period as a resting stage (Al Yamani et al. 2011).

Comments Unlike all other species, *L. simplex* showed its highest abundance during monsoon and early postmonsoon and differs from *L. nordquisti* in its absence of the aboral funnel and longer lorica. Its maximum abundance reached up to 450 ind l^{-1} in November 09 in Dhamakhali (S$_2$) of Indian Sundarbans with 26.7 °C temperature and 5.4 p.s.u salinity.

2.4.26 *Leprotintinnus norqvisti (Kofoid and Campbell, 1929)*

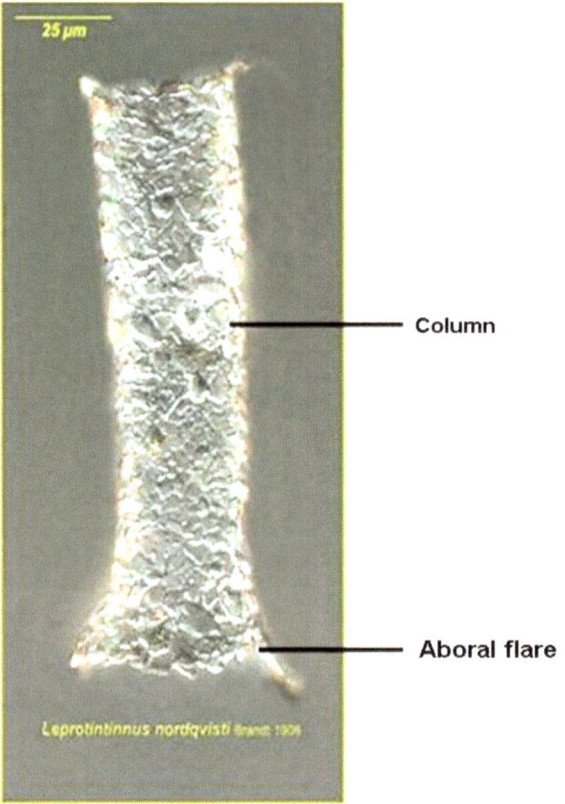

Phylum: Ciliophora
Subphylum: Intramacronucleata
Class: Spirotrichea
Subclass: Choreotrichia
Order: Tintinnida
Family: Tintinnidiidae

Morphological characteristics Lorica consisting of a tubular shaft and an inverted funnel-shaped aboral flare, oral rim irregular, usually slightly flaring; shaft more or less tapering, expanding near the posterior region to form a distinct aboral conical flare; aboral margin very ragged; wall showing a faint spiral structure, made of rather scare particles aggregated more thickly on the surface of the aboral flare than on that of the shaft.

Measurements: Length ranged from 54.89 to 59.21 μm and the LOD ranged from 24.66 to 30.41 μm.

Distribution Common in Kuwait waters and around Bubiyan Island. This species occurred along the Iranian coast with southeastern direction (Yamani et al. 2011) and from Southeast coast of India (Parangipettai) (Godhantaraman 2002).

Comments Distribution and abundance of this species are very much related to the abundance of phytoplankton of that particular site. Generally, it is a stenothermal and stenohaline species (Godhantaraman and Uye 2003). Maximum abundance (100 ind l^{-1}) was recorded from Gangasagar (S_5) of Indian Sundarbans during February 2010.

2.4.27 *Tintinnidium primitivum (Busch, 1923)*

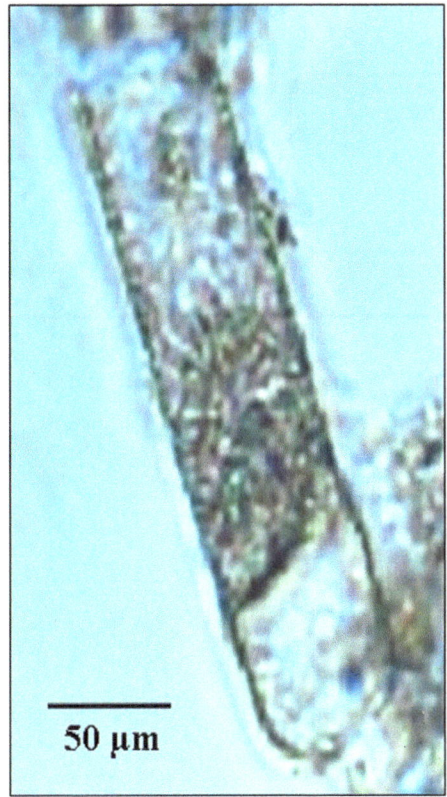

50 μm

Phylum: Ciliophora
Subphylum: Intramacronucleata
Class: Spirotrichea
Subclass: Choreotrichia
Order: Tintinnida
Family: Tintinnidiidae

Morphological characteristics Lorica rod shaped with an open aboral end. Soft and sparse agglomerated particles were included inside lorica.

Measurement The length of the lorica varied widely ranging from 25.82 to 37.39 μm and the oral diameter ranged from 8.09 to 12.67 μm. Highest lorica length (57.61 μm) was found in premonsoon (March 2010) followed by postmonsoon (Av: 36.72 ± 10.3 μm) and monsoon (Av: 25.2 ± 9.73 μm).

Distribution Recorded from Gulf of Mexico (Federal and Camp 2010) and Pichavaram mangrove, India (Godhantaraman 1994).

Comments This species was mostly frequent during postmonsoon season, but the maximum abundance 250 ind l^{-1}) was recorded during May 2010 at the site Canninig (S_3) of Indian Sundarbans when the temperature was 30.5 °C and the salinity was 25.8 p.s.u.

2.4.28 *Tintinnidium mucicola (Claparède and Lachmann, 1858) Daday, 1887*

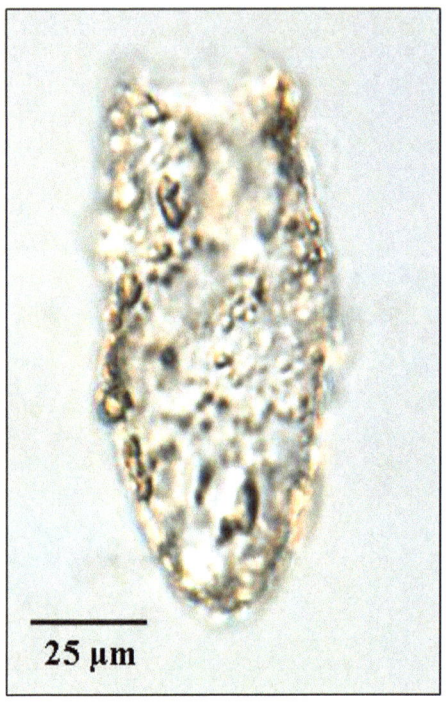

Phylum: Ciliophora
Subphylum: Intramacronucleata
Class: Spirotrichea
Subclass: Choreotrichia
Order: Tintinnida
Family: Tintinnidiidae

Morphological characteristics Lorica was not divided into collar and bowl. It is large, conical shaped with a blunt end. Agglomeration is sparse. No aboral horn present.

Measurement Body length varied from 30.58 to 37.48 μm and the LOD ranged from 9.82 to 11.50 μm. Ratio between length and oral diameter was ~1.0:3.25.

Distribution Found from Gulf of Mexico (Federal and Camp 2010) and Ekrainian exclusive economic zone (Alexandrov and Korshenko 2007).

Comments It was a typical postmonsoon species recorded maximum abundance (100 ind l^{-1}) during December 2010 at the site Dhamakhli of Indian Sundarbans. It prefers temperature range <25 °C and salinity between 13 and 15 p.s.u.

2.4.29   *Codonellopsis schabi (Brandt, 1906) Kofoid and Campbell, 1929*

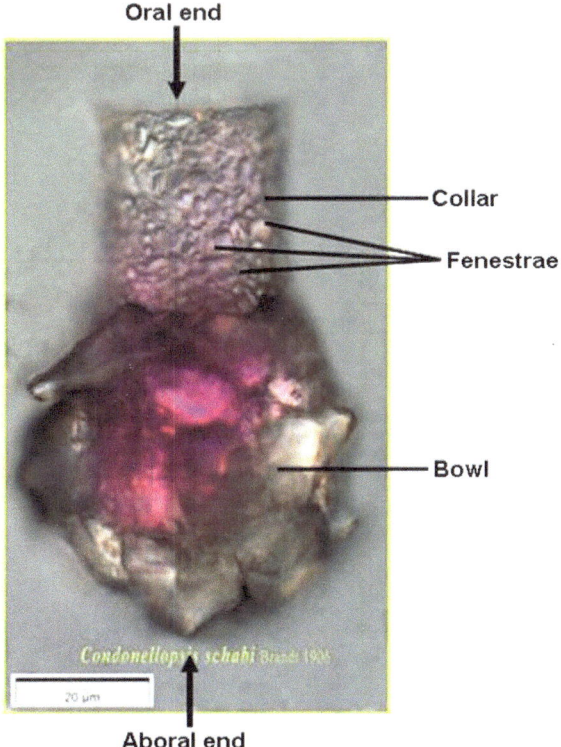

Phylum: Ciliophora
Subphylum: Intramacronucleata
Class: Spirotrichea
Subclass: Choreotrichia
Order: Tintinnida
Family: Codonellopsidae

Morphological characteristics Lorica divided into a collar and a medium-sized bowl; collar with a somewhat flaring rim, slightly bulging near its middle, usually lower than the bowl, its height 0.22–0.34 of the total length, composed of 4–11 spiral turns with a few elliptical fenestrae; bowl generally ovate, aboral region convex conical; aboral end usually round or rarely bluntly pointed; wall of the bowl thick, coarsely agglomerated.

Measurements Length and oral diameter varied from 65 to 95 μm and 29 to 32 μm, respectively. Length of the collar was 14–32 μm; greatest diameter of the bowl 43–65 μm. Approximate ratio between length and oral diameter was 2.2–3.2.

Distribution *Codonellopsis schabi* is common in the northwestern and central regions of the Arabian Gulf and in the Strait of Hormuz, with high abundance in the central region of the Gulf (Al Yamani et al. 2011), Gulf of Mexico (Federal and Camp 2010), and India (Godhantaraman 2002).

Comments Occasional species in Sundarbans mangrove wetland found only once during March 2010 associated with a monospecific bloom of *T. beroidea.*

2.4.30 *Stenosemella ventricosa (Claparède and Laachmann, 1858) Jörgensen, 1924*

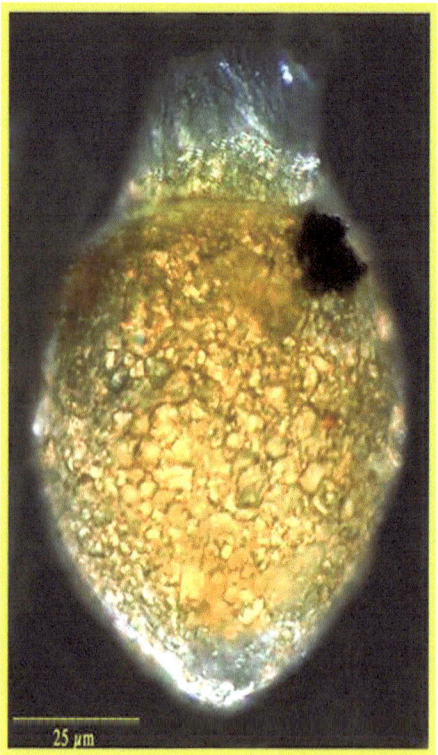

25 μm

Phylum: Ciliophora
Subphylum: Intramacronucleata
Class: Spirotrichea
Subclass: Choreotrichia
Order: Tintinnida
Family: Codonellopsidae

Morphological characteristics Lorica was divided into two parts—Collar and bowl both are short in length. Gutter absent between bowl and collar.

Measurement Lorica length found to be ~24.14 μm, while the LOD is also of almost similar size (23.7 μm).

Distribution Recorded from Gulf of Mexico (Federal and Camp 2010), Rijeca Bay (Zavodnik and Covacic 2000), Russian part of the black sea, Ukrainian exclusive economic zone Bulgaria (Alexandrov and Korshenko 2007) and Southeast coast of India (Godhantaraman 2002).

Comments Appeared mostly during postmonsoon period followed by monsoon when the temperature was between 11 and 16 °C. Although recorded in salinity values ~9 p.s.u, the abundance was maximum >17 p.s.u. It had an enlargement of the oral region formed by small particles attached to the hyaline collar. Maximum abundance (82 ind l^{-1}) was recorded during January 2011 at Dhamakhali of Indian Sundarbans.

2.4.31   *Favella ehrenbergii (Claparède and Laachmann, 1858)*

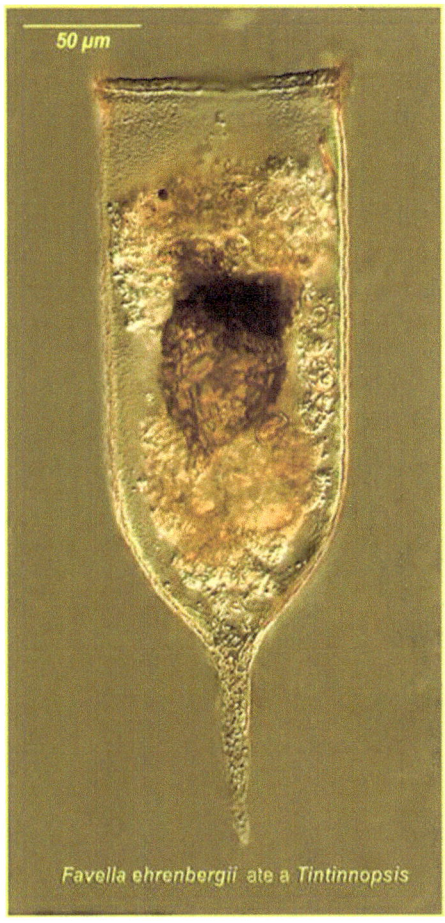

50 μm

Favella ehrenbergii ate a Tintinnopsis

Phylum: Ciliophora
Subphylum: Intramacronucleata
Class: Spirotrichea
Subclass: Choreotrichia
Order: Tintinnida
Family: Xystonellidae

Morphological characteristics Lorica long and cylindrical, wall thick, bowl sometimes slightly expanded below middle, rounded below and joined by wings to a short blunt, pedicel. Spiral turns were present sometimes suborally.

Reported measurements Total length 145–400 μm. Oral diameter 54–124 μm. Approximate ratio LOD 2.4–4.2.

Distribution Wide species recorded from Gulf of Mexico (Federal and Camp 2010), Rijeca Bay (Zavodnik and Covacic 2000), Romanian exclusive economic zone, Russian part of the black sea, Ukrainian exclusive economic zone Bulgaria (Alexandrov and Korshenko 2007), Southeast coast of India (Godhantaraman 2002) and Cochin back waters (Jyothibabu et al. 2008), India. But it was registered in low abundance in the northwestern waters of Arabian Gulf (Yamani et al. 2011).

Comments This species owned the largest LOD among all, and the existence of the second tintinnid inside *F. ehrenbergii* in the figure proves that it feeds not only upon phytoplankton but on smaller tintinnid also. This could be considered as an occasional species recorded only two times during the entire study period, coinciding with algal blooms in coastal regions of Sundarbans.

2.4.32 *Amphorellopsis acuta (Schmidt, 1901)*

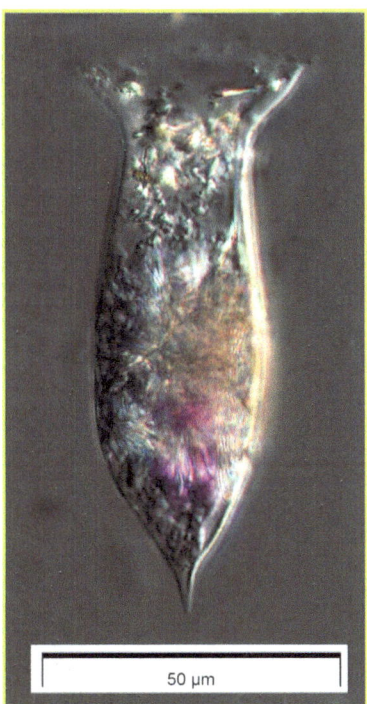

Phylum: Ciliophora
Subphylum: Intramacronucleata
Class: Spirotrichea
Subclass: Choreotrichia
Order: Tintinnida
Family: Tintinnidae

Morphological characteristics Lorica fusiform, oral aperture circular; collar low-funnel-shaped, bowl circular in cross section below the collar, then gradually becoming triangular, posteriorly with three ridges in the aboral end; wall hyaline, composed of separated laminae in the anterior of the lorica.

Measurements Length 90.98–93.16 μm; oral diameter 31.68–37.94 μm. Approximate ratio Lorica/oral diameter 2.3–3.9.

Distribution Common at all studied areas of the Arabian Gulf (Yamani et al. 2011), Gulf of Mexico (Felder and Camp 2010), Southeast coast of India (Godhantaraman 2002), Bahuda estuary, East coast of India (Mishra and Panigrahy 1999), Porto Novo region, India (Krishnamurthy and Santhanam 1975), East Asian waters (Lee and Kim 2000).

Comments This is a typical stenothermal and euryhaline species (Godhantaraman and Uye 2003). Mostly, thermophilic was found in water temperature >25 °C. It was found to tolerate a wide range of salinity (14.49–21.3 p.s.u) but showed highest density at >20 p.s.u.

2.4.33 Eutintinnus sp. (Kofoid and Campbell, 1939)

Phylum: Ciliophora
Subphylum: Intramacronucleata
Class: Spirotrichea
Subclass: Choreotrichia
Order: Tintinnida
Family: Tintinnidae

Morphological characteristics Lorica very small, almost cylindrical. Ciliate is bigger in comparison with lorica. Its body takes all the width of lorica and occupies about 80 % of its total length. The wall of lorica is very thin, transparent, and flimsy. In general, most of these tintinnids in the collected samples have more or less misshapen loricas.

Measurement Total length 53.4 μm; oral diameter 13.8 μm.

Distribution *Eutintinnus* sp. was registered near the north coast of Bubiyan Island and in Kuwait Bay.

Comments *Eutintinnus* sp. was registered in estuarine site Lot no. 8 of Indian Sundarbans only with abundance of 10 ind l^{-1} during postmonsoon (November 2012), when the temperature and salinity were 25 °C and 15.04 p.s.u, respectively.

2.4.34   *Metacylis mediterranea (Mereschkowsky, 1880) Jörgensen, 1924*

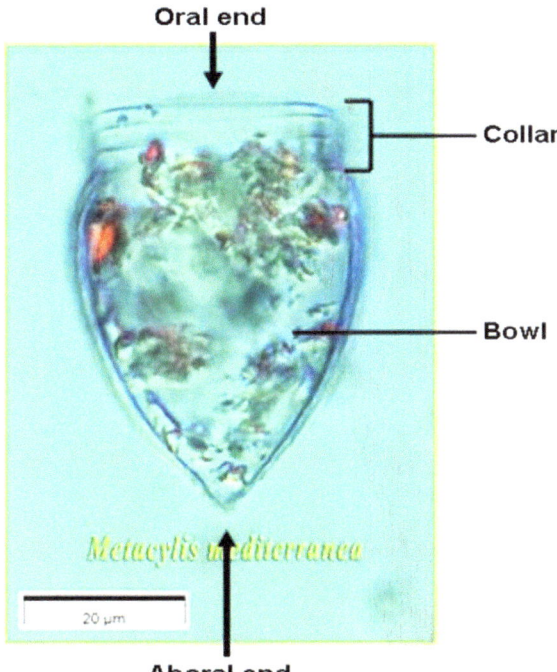

Phylum: Ciliophora
Subphylum: Intramacronucleata
Class: Spirotrichea
Subclass: Choreotrichia
Order: Tintinnida
Family: Metacylididae

Morphological characteristics Lorica small, hyaline, cup-shaped with rings or spiral bands on part or entire lorica. Aboral end is blunt and round. Collar wide, relatively short, with 4–6 spiral turns.

Measurement Total length of the lorica ranged from 68.11 to 70.75 μm and the oral diameter ranged from 38.41 to 41.60 μm. Approximate ratio between length and oral diameter was 1.0:1.7

Distribution Recorded mostly from Russian part of the Black Sea, Ukrainian exclusive economic zone Bulgaria (Alexandrov and Korshenko 2007), Porto Novo region, India (Mishra and Panigrahy 1999).

Comments Scarce species recorded only once in March 2010 at Lot 8 (S$_4$) of Indian Sundarbans. The temperature and salinity at that specific time were 32.85 °C and 21.3 p.s.u., respectively.

2.4.35 *Metacyllis sp. (Jörgensen, 1924)*

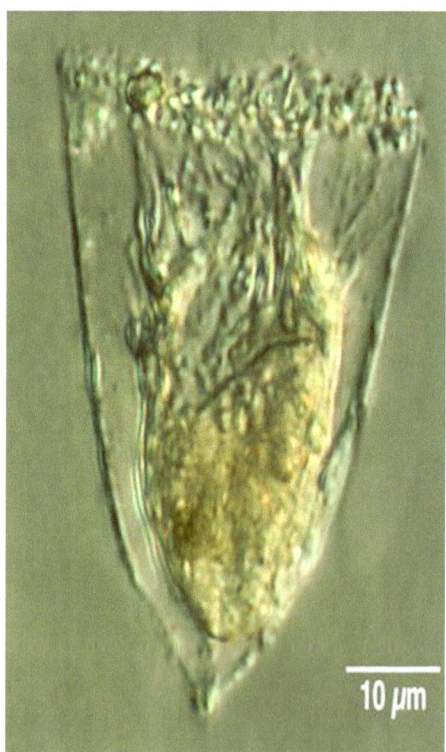

Phylum: Ciliophora
Subphylum: Intramacronucleata
Class: Spirotrichea
Subclass: Choreotrichia
Order: Tintinnida
Family: Metacylididae

Morphological characteristics Lorica small, hyaline, cup-shaped, with rounded aboral end. Collar wide, relatively short, with 4–6 spiral turns.

Measurements Length 65.55 μm; oral diameter 37.25 μm.

Distribution Common in Bubiyan Island waters, Kuwait.

Comments *Metacylis* sp. was registered only in Lot no. 8 (S$_4$) of Indian Sundarbans in very low abundance (50 ind l^{-1}) during premonsoon (April 2012).

2.4.36   *Helicostomella sp. (Jörgensen, 1924)*

Phylum: Ciliophora
Subphylum: Intramacronucleata
Class: Spirotrichea
Subclass: Choreotrichia
Order: Tintinnida
Family: Metacylididae

Morphological characteristics Lorica very short, bullet-shaped, 2.2–4.4 oral diameters in length; oral rim entire; bowl expanding slightly, widest 0.5–1.0 oral diameter below the spiral, convex conical aborally; aboral horn scarcely differentiated, conical.

Measurement Total length 57.95 μm; length of collar 16.4 μm.

Distribution *Helicostomella longa* was registered in very low abundance in the northwestern waters of the Arabian Gulf.

Comments *Helicostomella* sp. remarkably differs from other tintinnids in having shorter, stouter lorica. *Helicostomella* sp. was registered only in Lot no. 8 of Indian Sundarbans, with an abundance of 15 ind l^{-1}. It is an occasional species, found only during December 2012, coinciding with moderate temperature (25 °C) and salinity (15–17 p.s.u.).

2.4.37 *Wangiella dicollaria (Nie, 1934)*

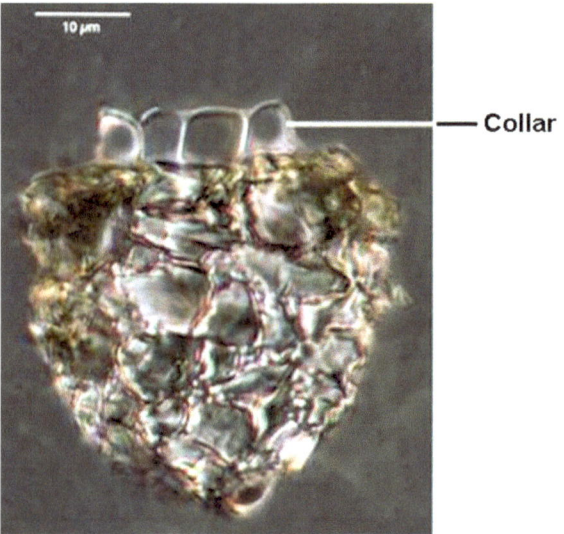

Phylum: Ciliophora
Subphylum: Intramacronucleata
Class: Spirotrichea
Subclass: Choreotrichia
Order: Tintinnida
Family: Dictyocystidae

Morphological characteristics Lorica with oral margin undulating, with a single row of 4–5 tall rectangular windows; beams sometimes bowed outwards. Bowl subglobular, or rarely wider than long, set off by distinct shoulder from the bowl; aboral region sub hemispherical. Wall of bowl with large, subuniform, coarse, rounded and overlapping meshes and with smaller ones intermingled, formed by included coccoliths.

Measurement Length 55.6 μm; oral diameter 36.3 μm.

Distribution   Recorded from Bohai Sea, China (Zhang et al. 2002).

Comments   The species differs from other representatives belonging to the family Dictyocystidae in the extensive coccolith inclusions and duplex nature of the wall structure. *Wangiella dicollaria* is rarely available in estuarine waters and recorded exclusively at the site Gangasagar (S$_5$) of Indian Sundarbans with an abundance of 20 ind l^{-1} during February 2013, coinciding with 20 °C temperature and 20 p.s.u salinity.

References

Al Yamani, F. Y., Skryabin, V., Gubanova, A., Khvorov, S., & Prusova, I. (2011). *Marine zooplankton practical guide for the Northeastern Arabian Gulf* (1st ed., vol. I). Kuwait Institute of Scientific Research.

Balech, E. (1951). Nuevos datos sobre Tintinnoinea de Argentina y Uruguay. *Physis, Buenos Aires, 20*, 291–302.

Bhattacharya, A., & Sarkar, S. K. (2003). Impact of overexploitation of shellfish: North eastern coast of India. *Ambio, 32*(1), 70–75.

Capriulo, G. M., & Carpenter, E. J. (1983). Abundance, Species Composition and Feeding Impact of Tintinnid Micro-Zooplankton in Central Long Island Sound. *Marine Ecology Progress Series, 10*, 277–288.

Clarke, K. R., & Warwick, R. M. (1994). *Change in marine communities* (144 pp). Plymouth: Plymouth Marine Laboratory.

Cosper, T. C. (1972). The Identification of Tintinnids (Protozoa: Ciliata: Tintinnida) of the St. Andrew Bay System, Florida. *Bulletin of Marine Science, 22*(2), 391–418.

De Pauw, N. (1975). Kwalitatieve en kwantitatieve analyse van het micro- en nannoplankton in de Belgische kustwateren ter hoogte van Nieuwpoort. In: F. Lievens (Ed.), *Ekologische en biologische studie van de kustwateren ter hoogte van Nieuwpoort in verband met het lozen van afvalwateren. C.L.O.* (pp. 75–203). Gent: Gent, Belgium.

Dolan, J. R., Claustre, H., Carlotti, F., Plounevez, S., & Moutin, T. (2002). Microzooplankton diversity: Relationship of tintinnid ciliate with resources, competitors and predators from the Atlantic Coast of Morocco to the Eastern Mediterranean. *Deep Sea Research, Part I, 49*, 1217–1232.

Federal, D. L., & Camp, D. K. (2010). Waters and Biota. Biodiversity. Texas A&M Press, College Station, Texas.

Felder, D. L., & Camp, D. K. (eds.) (2010). Gulf of Mexico: Origins, Waters, and Biota: Volume 1, Biodiversity (Texas: A&M University Press).

Godhantaraman, N. (1994). Species composition and abundance of tintinnids and copepods in the Pichavaram mangroves (South India). *Ciencias do Mar, 20*, 371–391.

Godhantaraman, N. (2001). Seasonal variations in taxonomic composition, abundance and food web relationship of microzooplankton in estuarine and mangrove waters, Parangipettai region, southeast coast of India. *Indian Journal of Marine Science, 30*, 151–160.

Godhantaraman, N. (2002). Seasonal variations in species composition, abundance, biomass and estimated production rates of tintinnids at tropical estuarine and mangrove waters, Parangipettai, southeast coast of India. *Indian Journal of Maine Sciences, 36*, 161–171.

Godhantaraman, N., & Uye, S. (2003). Geographical and seasonal variations in taxonomic composition, abundance and biomass of microzooplankton across a brackish-water lagoonal system of Japan. *Journal of Plankton Research, 25*, 465–482.

Hayward, P. J., & Ryland, J. S. (Ed.). (1990). The marine fauna of the British Isles and North-West Europe: *Introduction and protozoans to arthropods* (vol. 1). Oxford, UK: Clarendon Press. ISBN 0-19-857356-1. (p. 627).

Jiang, Y., Yang, J., Al-Farraz, S. A., Warren, A., & Lin, X. (2012). Redescriptions of three tintinnid ciliates, Tintinnopsis tocantinensis, T. radix, and T. cylindrica (ciliophora, spirotrichea), from coastal waters off China. *European Journal of Protistol, 48*(4), 314–325.

Jyothibabu, R., Madhu, N. V., Maheswaran, P. A., Asha Devi, C. R., Balasubramanian, T., & Nair, K. K. C. (2006). Environmentally related of symbiotic associations of heterotrophic dinoflagellates with cyanobacteria in the Bay of Bengal. *Symbiosis, 42*, 51–58.

Jyothibabu, R., Madhu, N. V., Maheswaran, P. A., Jayalakshmi, K. V., Nair, K.K.C., & Achuthankutty, C.T. (2008). Seasonal variation of microzooplankton (20-200 μm) and its possible implications in the western Bay of Bengal. *Continental Shelf Research, 28*, 737–755.

Kofoid, C. A., & Campbell, A. S. (1929). A conspectus of the marine and freshwater Ciliata belonging to the sub-order Tintinnoinea with descriptions of new species principally from the Agassiz Expedition to the eastern tropical Pacific, 1904–1905. *University of California Publications in Zoology, 34*, 1–403.

Kofoid, C. A., & Campbell, A. S. (1939). Reports on the scientific results of the expedition to the Eastern tropical Pacific. The Ciliata: The Tintinnoinea. *Bulletin of the Museum of Comparative* Zoology, *Harvard, 84*, 1473.

Krishnamurthy, K., & Santhanam, K. (1975). Ecology of tintinnids (Protozoa: Ciliata) in Porto Novo region. *Indian Journal of Maine Sciences, 41*, 181–184.

Marshall, S. M. (1969). Protozoa order tintinnida. In J. H. Fraser & V. Kr. Hansen (Eds.), *Fishes d' Identification du Zooplankton Cons. Per. Int. Explor. Mer Zooplankton Sheet* (pp. 112–117). Charlottenlund, Denmark.

Mishra, S., & Panigrahy, R. C. (1999). The Tintinnids (Protozoa: Ciliata) of the Bahuda estuary, east coast of India. *Indian Journal of Maine Sciences, 28*, 219–221.

Mishra, S., Sahu, G., Mohanty, A. K., Singh, S. K., & Panigrahy, R. C. (2006). Impact of the diatom *Asterionella glacialis* (Castracane) bloom on the water quality and phytoplankton community structure in coastal waters of Gopalpur sea, Bay of Bengal. *Asian Journal of Water Environment Pollution, 3*, 71–77.

Müller, H., & Geller, W. (1993). Maximum growth rates of aquatic ciliate protozoa: The dependence on body size and temperature reconsidered. *Archiv fuer Hydrobiologie, 126*, 315–327.

Muller, Y. (2004). Faune et flore du littoral du Nord, du Pas-de-Calais et de la Belgique: Inventaire. [Coastal fauna and flora of the Nord, Pas-de-Calais and Belgium: inventory] (p. 307). Commission Régionale de Biologie Région Nord Pas-de-Calais: France.

Naha Biswas, S., Godhantaraman, N., Rakshit, D., & Sarkar, S. K. (2013). Community composition, abundance, biomass and productive rates of Tintinnids (Ciliata: Protozoa) in the coastal regions of Sundarbans Mangrove wetland, India. *Indian Journal of Geo- Marine Sciences, 42*(2), 163–173.

Naskar, K. R., & Guha Bakshi, D. N. (1987). *Mangrove swamps of the Sundarbans—An ecological perspective* (264 p). Calcutta, India: Naya Prakash.

Nie, D. S. (1934) Notes on tintinnoinea from the Bay of Amoy. Marine Biological Association of China, Third Annual Report: 71–80.

Selvan, V. (2003). Environmental classification of mangrove wetlands of India. *Current Science, 84*, 12–20.

Sitran, R., Bergamasco, A., Decembrini, F., & Guglielmo, L. (2009). Microzooplankton (tintinnid ciliates) diversity: coastal community structure and driving mechanisms in the southern Tyrrhenian Sea (Western Mediterranean). *Journal of Plankton Research, 31*, 153–170.

Sokal, R. R., & Rohlf, J. F. (1981). *Biometry: The principles and practice of statistics in biological research* (2nd ed., 859 p). San Francisco: W. H. Freeman and Company.

Stanley, D. J., & Hait, A. (2000). Holocene depositional patterns, neotectonics and Sundarbans mangroves in the Western Ganges-Brahmaputra Delta. *Journal of Coastal Research, 16*, 26–34.

Strickland, J. D. H., & Parsons, T. R. (1972). *Practical handbook of sea water analysis* (310 pp). Canada, Ottawa: Fish Research Board (Bulletin No. 167).

Urrutxurtu, I. (2004). Seasonal succession of tintinnids in the Nervion River estuary, Basque Country, Spain. *Journal of Plankton Research, 26*, 307–314.

Verity, P. G., & Langdon, C. (1984). Relationship between lorica volume, carbon, nitrogen, and ATP content of tintinnids in Narragansett Bay. *Journal of Plankton Research, 66,* 859–868.

Von Daday, E., (1887). Monographie der Familie der Tintinnodeen. Mitt. zool. Stn. Neapel., 7:473–591.

Xu, H., Song, W., Warren, A., Al-Rasheid, K. A. S., Al-Farraj, S. A., Gong, J., et al. (2008). Planktonic protist communities in a semi-enclosed mariculture pond: Structural variation and correlation with environmental conditions. *Journal of the Marine Biological Association, UK, 88,* 1353–1362.

Zavodnik, D., & Kovacic, M. (2000). Index of marine fauna in Rijeka Bay (Adiartic sea, Croatia). *Natural Croatica, 9*(4), 297–379.

Zhang, W., Xu, K., Wan, R., Zhang, G., Meng, T., Xiao, T., Wang, R., Sun, S., Choi, J. K. (2002) Spatial distribution of ciliates, copepod nauplii and eggs, Engraulis japonicas post-larvae and microzooplankton herbivorous activity in the Yellow Sea, China. *Aquatic Microbial Ecology, 27,* 249–259.

Chapter 3
Results and Discussion

Abstract This chapter presents the in-depth case study dealing with the community composition, numerical abundance, biomass, and production rate of the ciliated tintinnids in the coastal waters of Sundarbans mangrove wetland for two consecutive years (2009–2011). A total of 37 species (31 agglomerated and 6 non-agglomerated) belonging to 11 genera have been identified from five study sites in Indian Sundarbans where *Tintinnopsis* (24 spp) represented the most dominant genera. Pronounced seasonal variations were noticed in environmental parameters and abundance, biomass, production rates of tintinnids. Water temperature, chl *a*, and salinity were found to regulate the seasonal tintinnid variations as revealed by stepwise multiple regression analyses. The changes in lorica morphology of a tintinnid in terms of temperature and salinity are discussed.

Keywords Tintinnids · Community structure · Key species · Biomass and production rate · Seasonal cycle · Polymorphism

3.1 Environmental Variables

3.1.1 Dhamakhali (S_1)

An overall trend of high temperature was noticed at this site, ranging from 22.85 to 32.7 °C. Due to the largest distance from the sea, salinity range was lower in comparison with all other study sites of Sundarbans ranging from 5.4 to 20.5 p.s.u with a mean value of 15.13 ± 5.75 p.s.u. A medium range of dissolved oxygen ($4.77–6.74$ mg l^{-1}) and high turbidity value (Av: 16.40 ± 3.65 NTU) were recorded throughout the year which also affected the total tintinnid community structure. Surface water pH ranged from 8.0 to 8.36 (mean 8.19 ± 0.11) where the maximum value was recorded during March 2011 and the minimum was recorded during August, 2010. High turbidity of the surface water restricts the light penetration which in turn retards the generation of plankton population. Nutrients

© The Author(s) 2015
S.K. Sarkar, *Loricate Ciliate Tintinnids in a Tropical Mangrove Wetland*,
SpringerBriefs in Environmental Science, DOI 10.1007/978-3-319-12793-4_3

were always found in high concentrations in this site. The level of nutrients in the rivers is influenced by natural factors such as catchment geology, rainfall, and river flow patterns. However, land use also has a large influence. In this case, a canal which runs directly from Calcutta passes through Dhamakhali and ultimately gets mixed with the water of Bhagirathi. As a result, all the industrial as well as domestic sewages carried by this canal get mixed with the water of Bidyadhori River in Dhamakhali and subsequently increases the concentration of nutrients. Both nitrate (61.19 µg atm l^{-1}) and silicate (93.48 µg atm l^{-1}) showed their maximum concentration during postmonsoon period in November 09 and November 10, respectively. Chl a ranged from 0.67 to 2.55 mg l^{-1} with a mean value of 1.43 $\pm$ 0.62 mg m^{-3}.

3.1.2 Canning (S_2)

Surface water temperature ranged between 19.0 and 32.65 °C throughout the year as summarized in Table 3.1. Salinity also showed a wide variation (8.10–25.85 p.s.u) with a mean value of 17.33 $\pm$ 6.81 p.s.u. This site is comparatively sheltered with a very prominent tidal effect. Turbidity did not show much variation ranging between 7 and 16.5 NTU. Maximum turbidity (16.5 NTU) was recorded during September; 2010. Comparatively lower turbidity helped the plankton community structure as well as the chl a to increase their concentrations. Low pH range was found throughout the year ranging from 7.24 to 7.96 where the maximum value was recorded during May 2011 and the minimum during February, 2010. Overall, high D.O. range was found during the study period with the mean value of 5.59 $\pm$ 1.32 mg l^{-1}. Nutrients showed increasing trend during early postmonsoon period when the monsoonal runoff got mixed with the water. All the three nutrients, viz. Nitrate (25.75 µmol l^{-1}), phosphate (1.5 µmol l^{-1}), and silicate (73.7 µmol l^{-1}) showed their maximum concentration during October, 2011. Chl a exhibited maximum mean value (2.34 $\pm$ 1.3 mg m^{-3}) in this sheltered site infested with mangroves.

3.1.3 Lot 8 (S_3)

Surface water temperature ranged from 21.65 to 32.85 °C with an overall higher trend as summarized in Table 3.1. Among all the sites, Lot 8 (S_5) showed maximum mean temperature (29.3 $\pm$ 3.15 °C). Maximum salinity (22 p.s.u.) was found during the summer months (April, 2011), and the minimum (6.0 p.s.u.) was found during the monsoon months (August, 2010). An overall high turbidity (Av: 14.34 $\pm$ 4.93) and low dissolved oxygen (Av: 5.33 $\pm$ 0.91 mg l^{-1}) value was recorded at this site throughout the year which might be due to high siltation. The correlation coefficient (r) between turbidity and temperature showed a positive

Table 3.1 Pooled mean values of the water quality parameters at five sampling sites (S_1–S_5)

Environmental parameters	Sampling sites				
	Dhamakhali S_1	Canning S_2	Lot 8 S_3	Chemaguri S_4	Gangasagar S_5
Temp (°C)	28.55 ± 3.57	28.12 ± 4	29.3 ± 3.15	28.51 ± 3.58	28.96 ± 4.02
Salinity (p.s.u.)	15.13 ± 5.75	17.33 ± 6.81	12.25 ± 5.95	17.42 ± 5.72	22.70 ± 6.39
Turbidity (NTU)	16.40 ± 3.65	11.08 ± 3.39	14.34 ± 4.93	17.15 ± 3.67	12.53 ± 4.23
pH	8.19 ± 0.11	7.75 ± 0.2	8.25 ± 0.26	8.15 ± 0.33	8.24 ± 0.19
D.O. (mg l^{-1})	5.72 ± 0.51	5.59 ± 1.32	5.33 ± 0.91	5.32 ± 0.56	5.43 ± 0.55
B.O.D. (mg l^{-1})	1.24 ± 0.33	1.11 ± 0.38	1.44 ± 0.47	1.11 ± 0.38	0.97 ± 0.46
Nitrate (μmol l^{-1})	37.52 ± 11.87	16.35 ± 4.49	17.68 ± 8.33	19.82 ± 6.13	14.46 ± 6.16
Phosphate (μmol l^{-1})	0.76 ± 0.39	0.63 ± 0.39	0.86 ± 0.59	1.47 ± 4.44	0.45 ± 0.63
Silicate (μmol l^{-1})	67.06 ± 16.63	45.79 ± 18.06	85.36 ± 34.61	75.77 ± 19	69.44 ± 24.28
Chl a (mg m^{-3})	1.43 ± 0.62	2.34 ± 1.3	2.20 ± 2.15	1.84 ± 1.61	2.25 ± 1.7

value ($r = 0.50$, $p = 0.01$) as turbidity is the condition resulting from suspended solids in the water, including silts, clays, industrial wastes, sewage, and plankton. Such particles absorb heat in the sunlight, thus raising water temperature, which in turn lowers dissolved oxygen levels. Both pH and D.O. showed a high value during bloom period where significantly low D.O. concentration was noticed during postbloom period. The concentration of nutrients (nitrate, silicate, and phosphate) showed a similar trend of temporal variations, with high concentration during monsoon and postmonsoon months (August–November) and low during the summer (March–May). This phenomenon could be attributed to monsoonal runoff from the adjacent agricultural fields, mangrove vegetation, and aquacultural firms. Maximum concentrations of chl a (3.57 mg m^{-3}) and numerical abundance of tintinnids (105 ind l^{-1}) were recorded during March 2010 from this site.

3.1.4 Chemaguri (S4)

Surface water temperature ranged from 21.0 to 33.5 °C whereas the mean salinity value was 17.42 ± 5.7 p.s.u accounting maximum value (29.3 p.s.u) in premonsoon (March, 2010). The minimum salinity value (7.5 p.s.u) was recorded during monsoon month (August, 2011) mainly due to freshwater discharge from the upstream regions as well as from the nearby agricultural fields and aquacultural firms. The surface water pH during the present study ranged from 8.0 to 8.7. Maximum (8.55) and minimum (7.3) values of pH were recorded during premonsoon (May, 2011) and postmonsoon (September, 2011), respectively. The maximum value might indicate high rate of illumination and subsequent photosynthesis (Mahajan and Kanhere 1995) and removal of CO_2. The decrease in pH during monsoon is due to the rainfall, land discharge runoff, and decrease in salinity (Marichamy et al. 1985). Concentration of D.O. ranged from 4.46 to 6.24 mg l^{-1} with a mean value of 5.32 ± 0.56 mg l^{-1} (as shown in Table 3.1). All the micronutrients (nitrate, phosphate, and silicate) showed the similar trend of temporal variation, high concentration during monsoon months, and lowest values during premonsoon (Table 3.1). The high concentration of nutrients in the surface water during monsoon and postmonsoon could be attributed to the monsoonal runoff from the agricultural fields, mangrove vegetation, and the aquacultural farms situated at the mouth of the creek. Chl a concentration ranged from 0.88 to 2.84 mg m^{-3} with a mean value of 1.8 ± 1.6 mg m^{-3}. This phytopigment proved itself as a crucial factor as most of the dominant tintinnid species showed significant positive correlation.

3.1.5 Gangasagar (S5)

The annual mean of water quality parameters showed an overall variation in their concentrations. Prevalence of high-saline conditions (15.4–30.8 p.s.u.) was observed throughout the year. The fluctuation of salinity from high and low values concurrent with non-monsoon and monsoon season (Menon et al. 1971). The lowest salinity was found during the monsoon months due to fresh water discharge from the upstream regions. The surface water pH during the present study ranged from 7.9 to 8.5. pH profile altered with seasons (Abhay Kumar and Dube 1995), Maximum value of pH was found in summer (March–May) and minimum in monsoon (June–August). The maximum pH recorded during May might be the indicative of high rate of illumination and subsequent photosynthesis (Mahajan and Kanhere 1995) and removal of CO_2 by photosynthesis (Kannan and Kannan 1996). The decrease in pH during June–August might be due to rainfall, land discharge runoff, and decrease in salinity (Marichamy et al. 1985). Concentration of D.O. in the surface water ranged from 4.7 to 6.38 mg l^{-1}. All the micronutrients (nitrate, silicate, and phosphate) showed the similar trend of temporal variation, high concentration during monsoon months (July–October), and lowest during premonsoon (March–May). Mean concentration of chl a varied from 0.81 mg m^{-3} (August) to 4.31 mg m^{-3} (November). It has been recorded that the concentration of chl b is always much less than chl a and c and this might be due to dominance of diatoms in this estuarine system (Mukhopadhyay et al. 2006), which mainly contain C_1 and C_2 as their phytopigments (Reynolds 2006).

3.2 Taxonomic Composition and Seasonal Species Distribution

The seasonal tintinnid community structure consists of 37 species dominated by 3 core species (present in substantial number almost throughout the year namely, *Tintinnopsis beroidea*, *T. minuta*, and *T. lohmani*), 16 seasonal and 18 occasional species. They were primarily grouped into agglomerated and non-agglomerated forms where the former group constituted the major part both qualitatively (31 sp) and quantitatively (75–100 % of the total community). Agglomerated form was absolutely dominated by the genera *Tintinnopsis* (24 sp) owing to its more flexible adaptive strategies (Aleya 1991; Reynolds 1997) followed by *Tintinnidium* (2 sp), *Stenosemella* (1 sp), *Leprotintinnus* (2 sp), *Codonellopsis* (1 sp), and *Wangiella* (1 sp). Non-agglomerated genera include *Favella* (1 sp), *Amphorellopsis* (1 sp), *Eutintinnus* (1 sp), *Metacylis* (2 sp), and *Helicostomella* (1 sp) and occupy 4–25 % of the total community. During November, 2010, an outburst of the centric diatom *H. hardmannianus* (along with other phytoplankton) was coincided with the proliferation of a solitary species *Favella ehrenbergii* and has raised the total abundance of tintinnid to a maximum (750 ind l^{-1}) at Gangasagar (S_5).

The major peaks of tintinnids coincided with a similar peak for diatoms and dinoflagellates. This result is not surprising, because ciliates feed largely on autotrophic nanoplankton organisms and small diatoms (Kimor 1969), which are very abundant during this period (Abboud-Abi Saab 1985). In addition, the peak density and diversity of tintinnid were found to be prominently accompanied by a high concentration of dinoflagellate and this coeval presence of dinoflagellate and tintinnids have been reported earlier by Lebour (1922). The positive relationship appears to be related to a nutritional interdependence between ciliates and the various groups of phytoplankton. Kimor and Golandsky (1977) noted such inter-dependence in the Red Sea. This relationship appears to be an important factor governing the population dynamics of tintinnids (Abboud-Abi Saab 1989).

3.2.1 Dhamakhali (S₁)

A total of nine tintinnid species were recorded from this site among which *Leprotintinnus simplex* was the most dominant one forming the bulk of the biomass (up to 78 %) followed by *T. beroidea* (up to 50 %). The distribution of *L. simplex* was exclusively found during monsoon and postmonsoon months with high concentrations absent or present in very low concentration during premonsoon. *T. beroidea* was the codominant species noticed almost through-out the year with comparatively low concentrations and thus may be considered as the eurythermal and euryhaline species. *T. minuta* was a premonsoon species tolerating high water temperature which was also endorsed by Capriulo and Carpenter (1983) from Central Long Island Sound. This species covered almost 15–28 % of the total tintinnid community during premonsoon period accounting maximum abundance (250 ind l^{-1}) during May, 2010. On the other hand, the lorica as well as the oral diameters of three postmonsoon species (*Tintinnopsis lobiancoi, Tintinnidium mucicola,* and *Stenosemella ventricosa*) were suffi-ciently high to cope up with the large-sized diatoms (*Chaetoceros lorenzianus, Hemidiscus hardmannianus, Thalasiosira decipiens, Coscinodiscus radiatus*) present during this time. Both *Tintinnopsis tubulosa* and *Tintinnopsis uruguay-ensis* may be considered as occasional species, encounter in very few cases throughout the study period.

3.2.2 Canning (S₂)

T. beroidea was the most dominant out of 13 tintinnid species forming bulk of the tintinnid community (56–100 %) followed by the codominant species *T. minuta* (41–50 %) and *Tintinnidium primitivum* (37–40 %). The pattern of bimodal type of tintinnid distribution closely followed that of biomass with relatively higher val-ues in premonsoon and a gradual decreasing trend toward monsoon and again ris-ing during postmonsoon period. Among the 13 species recorded, *T. beroidea* was found to be euryhaline and eurythermal as it was found almost throughout the year with varying concentration. The persistence of *T. beroidea* as the most abundant and frequently observed species in the coastal region of Sundarbans shows its unique capability to adjust to changing environmental conditions as endorsed by Modigh and Castoldo (2002). The species is considered a picoplaktivorous species and dominance of this agglutinated species seems to be related to the availability of particles to construct the lorica in addition to the presence of its preferred food.

The codominant species were replaced seasonally such as, *T. lobiancoi, T. lohmani, T. uruguayensis,* and *L. simplex* occurring ~26–51 % of the total tintinnid community in concerned seasons. The dominance of small-sized tintinnids, namely, *T. minuta* (15–28 %), *T. tubulosa* (4–15 %), *T. mucicola* (~15 %), and *T. lobiancoi* (5–10 %) which contribute up to 60 % of the total microzooplankton community, was evident at this site. This phenomenon seems to be a consequence of the high availability of bacteria and small flagellates (Urrutxurtu et al. 2003). It has been suggested that copepods selected large ciliates (>40 μm) over small ones (Nielsen and Kiorboe 1994; Vincent and Hartmann 2001) which may explain why small-sized ciliates dominated in this estuarine systems.

Rests of the species were occasional. A distinct seasonal variation in size of the oral diameter and lorica length was recorded from this site. The smallest lorica length (17–25 μm) was observed during postmonsoon (January) period when the water temperature was ~18 °C. During premonsoon, the lorica length became many fold higher (ranged between 80 and 110 μm) at a temperature of ~32 °C.

3.2.3 Lot 8 (S_3)

The tintinnid species ($n = 17$) showed similar distribution pattern like Canning (S_2), where *T. beroidea* formed the key species and form the bulk of the tintinnid community (38.9–100 %) based on their persistence and dominance. This species is followed by *L. simplex* which showed varying mode of distribution (up to 35 and 74 % during premonsoon and monsoon months, respectively) and *T. lohmani* which covers ~14–50 % of total tintinnid abundance. The pattern of tintinnid distribution showed relatively higher values during November–December with an increasing trend during premonsoon months (March–May). During March–April, the tintinnids showed maximum numerical abundance when the temperature was high. It is most surprising that out of 17 species recorded from this site, 8 are occasional and found only once throughout the study period. Among them, 6 species associated together during March 2010 with a "monospecific bloom" of *T. beroidea*.

3.2.4 Chemaguri (S_4)

Total 11 species belonging to 4 genera were identified from this site among which *L. simplex* could be considered as key species forming the bulk of the community (31–100 %) followed by *T. beroidea* (50–66 %) and *T. minuta* (47–50 %). The occurrence of *L. simplex* was exclusively confined during late monsoon months (August–September) when the species accounted 100 % of the total tintinnid community reflecting it preference to limnetic habitat and the significant negative correlation with salinity ($r = -0.68$; $p = 0.037$). At the same time, a positive correlation of this species with nutrients and turbidity was also established from the result of stepwise multiple regression analyses. Both the hydrological parameters were found in high concentrations during monsoon and late monsoon period because of input of fresh water from adjacent agricultural fields. Coinciding with these environmental conditions, the abundance of *L. simplex* also remained maximum during this period. Similar observation was reported by Godhantaraman (2002) from Parangipettai, southeast coast of India where *L. simplex* was found only during monsoon. Rest of the species was found occasional except *T. lobiancoi,* which is a typical postmonsoon species accounting ~35–50 % of the total community. According to Anna and Bjornberg (2006), *T. lobiancoi* is thermophilic in nature, occurring principally at low temperature (<25 °C) which endorses its presence during postmonsoon period. A significant variation between months ($F = 4.71, 0.001$) was also endorsed by ANOVA result.

3.2.5 Gangasagar (S_5)

A total of 10 species under 4 genera were identified from this site, most of them were occasional and were present only in few cases during the study period. An overall impoverished species diversity and density were noticed in this high-energy site reflected by the degree of variability in hydrological conditions. Apart from these, two large-sized tintinnid species, viz. *T. radix* (Length = 245.39–311.86 μm, L.O.D = 61.98–63.77 μm) and *Favella ehrenbergii* (Length = 292.76–322.5 μm, L.O.D = 92.45–95.72 μm), which never recorded before at this site, appeared and showed a huge abundance (200 and 83 ind l^{-1}, respectively) during November 2010 accompanied by a sudden outburst of phytoplankton community dominated by a centric diatom *H. hardmannianus*. It is noteworthy to mention that few species of tintinnids (only those that have large peristome and oral lorica diameters) are able to graze on phytoplankton during the bloom (Barría de Cao et al. 1997). Here, the dominating diatom species *H. hardmannianus* was found to be sufficient large (1, 17,068 μm^2 surface area) making a suitable environment for the large-sized tintinnid only. These two species subsequently disappeared from the pelagic system coinciding with the disappearance of *H. hardmannianus*.

3.3 Seasonal Variations of Tintinnid Abundance, Biomass and Production Rate

A wide seasonal variations in tintinnid abundance was observed with the maximum (~1,225 ind l^{-1}) during premonsoon (March–April) and the minimum (~75 ind l^1) during monsoon (August–September). Higher values during premonsoon months (May–June) might be attributed to elevated water temperature and chl *a* concentration, which are both considered as important factors for increased growth rate (Godhantaraman 2002). On the other hand, the lower abundance during monsoon period might be due to the non-conductive environment as observed by (Godhantaraman 1994) from Pichavaram mangrove, southeast coast of India. Species richness runs almost proportionally with species abundance for each month. Distinct seasonal succession in tintinnid community structure was very much pronounced in Sundarbans coastal region. Typical premonsoon species included *T. kofoidi*, *T. directa*, *T. cylindrica*, *T. radix*, and *T. parva*, However, *L. simplex*, being a typical monsoon species, was found in considerable numbers during postmonsoon. *T. gracilis*, *T. nucula*, *T. uruguayensis*, *T. tubulosa*, *T. mucicola*, *S. ventricosa*, *T. nana,* and *Helicostomella* sp. were noticed only during postmonsoon season. Most of these species (*T. uruguayensis*, *T. gracilis*, *T. mucicola*, and *S. ventricosa*) were confined only at two sheltered sites such as Canning (S_2) and Dhamakhali (S_1). *Amphorellopsis acuta* was recorded in both summer and winter months, and *F. ehrenbergii* was recorded once in association with severe proliferation of a diatom *H. hardmannianus*. Apart from this, *T. minuta* and *T. lohmani* also showed almost year round existence with the maximum concentration during summer months. Large number of smaller-sized tintinnids in summer months might be related to stable temperature and salinity regimes of the water. The dominance of smaller-sized food items like naked flagellates, diatoms, and peridinians present during this season could also be the reason for their peak abundance as ascertained by Godhantaraman (2002). Temporal changes in tintinnid population have been depicted in Fig. 3.1.

The biomass of the loricate ciliate was minimum during monsoon (0.06 µg Cl^{-1}) and maximum during summer months (4.81 µg Cl^{-1}). In general,

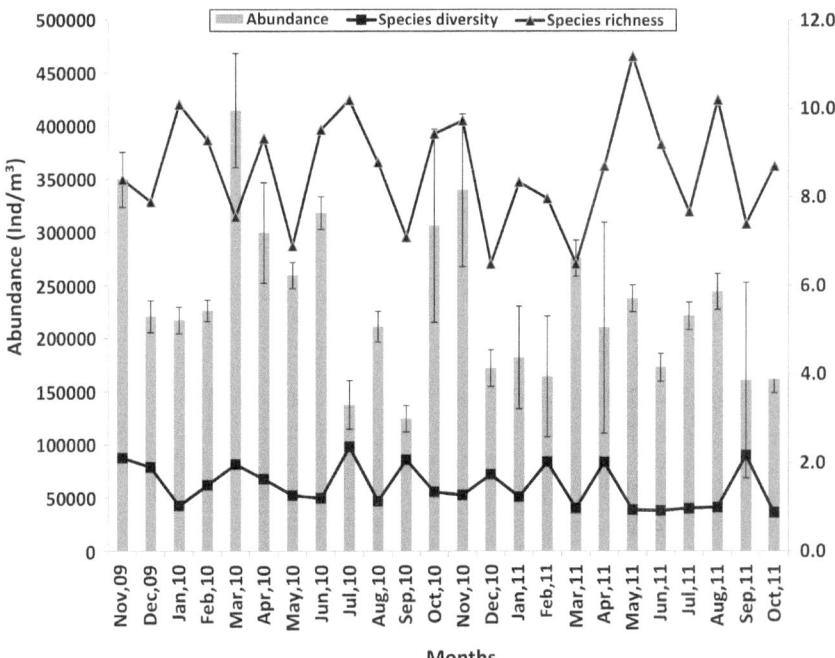

Fig. 3.1 Temporal changes in mean abundance, species diversity, and species richness of tintinnids in Sundarbans

majority of the species were of smaller lorica length (<65 μm) and their contribution to biomass was large. More or less similar biomass value was reported
by Godhantaraman and Uye (2001) from Parangipettai, southeast coast of
India. Maximum and minimum daily production rates ranged from 7.65 to
0.006 μg $Cl^{-1} day^{-1}$ during summer and monsoon season, respectively.

3.3.1 Dhamakhali (S₁)

Numerical abundance of tintinnids (450 ind l^{-1}) was recorded during November
2009 coinciding with the minimum salinity value (5.4 p.s.u) as shown in
Fig. 3.2a–c. The negative preference of this species for salinity was more strongly
established from the result of correlation coefficient ($r = -0.680$, $p = 0.015$) and
stepwise multiple regression analysis. The general high turbidity trend of this site
may also induce this species to flourish as it showed strong positive correlation
with turbidity ($r = 0.865$, $p = 0.00$). As the dominant species flourished mostly
during monsoon, total abundance of tintinnid at this site was also maximum at
that time only. In contrast, rest of the sites showed a uniform trend of maximum
abundance either in premonsoon or postmonsoon seasons subject to the water
quality characteristics of that site. The huge variation in the abundance of tintinnid between different months was also revealed by the result of ANOVA ($F = 3.2$,
$p = 0.005$). The abundance was shown in Fig. 3.2a–c.

   Biomass of a species varies with the abundance and size. As the maximum
abundance from Dhamakhali was recorded during postmonsoon followed by monsoon and premonsoon, biomass also followed the similar trend. The maximum
value of biomass was noticed during November 2009 and the lowest was noticed
during April 2010. Production rate was maximum (1.13) during October 2011 and
the lowest (0.04) was recorded during February 2010 (Fig. 3.3).

3.3.2 Canning (S₂)

The maximum concentration (250 ind l^{-1}) of this species was noticed during
March 2011 followed by November 09 (205 ind l^{-1}). *T. minuta* was also found
to be a premonsoon species in this site with maximum concentration (166 ind l^{-1})
found during April, 2010. Large number of smaller-sized tintinnids in summer
months might be related to stable temperature and salinity regimes of the water.
The dominance of smaller-sized food items like naked flagellates, diatoms, and
peridinians present during this season could also be the reason for their peak abundance as ascertained by Godhantaraman (1994). *T. primitivum* was a postmonsoon species with its maximum abundance (165 ind l^{-1}) found during November

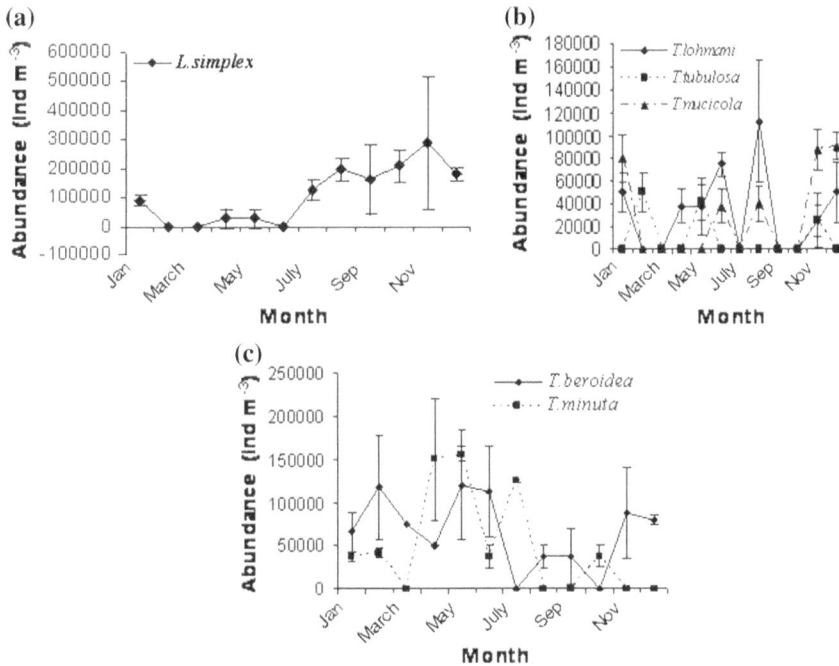

Fig. 3.2 a–c Temporal distribution of the dominant tintinnids species at Dhamakhali (S₁) (the annual mean and standard deviations are expressed in ind m⁻³)

Fig. 3.3 Seasonal variation of biomass and production rate at site S₁

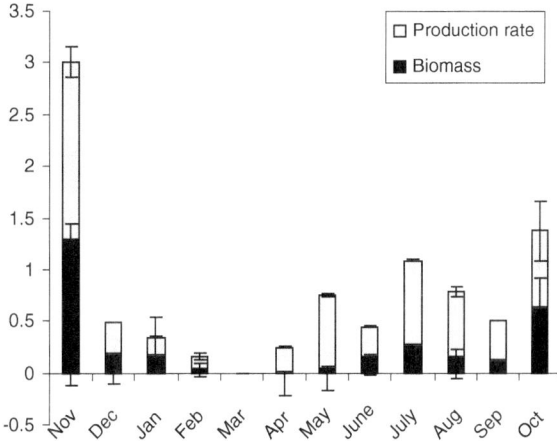

2010. However, this species was found to be highly correlated with the changing environment of postimmersion period after Durga puja in the upstream of river Ganges. The abundance was shown in Fig. 3.4a–c.

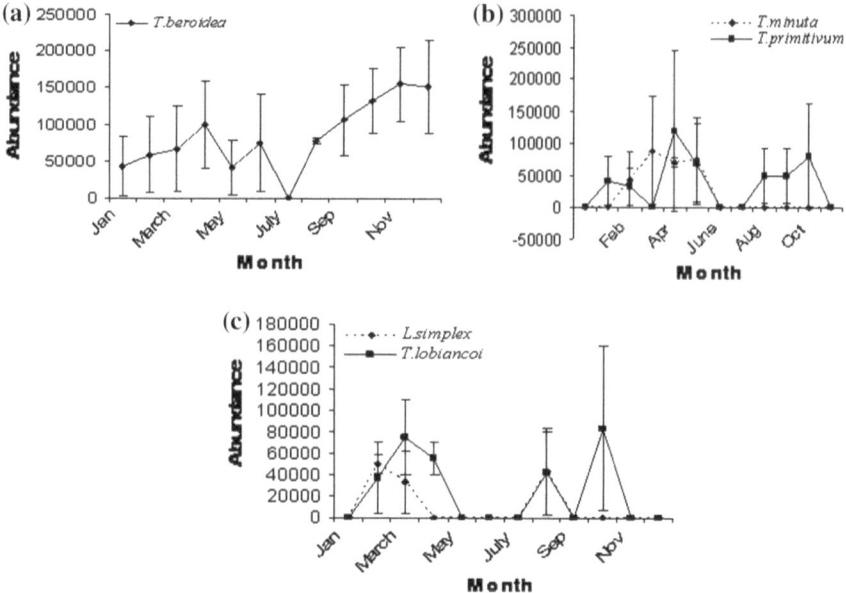

Fig. 3.4 a–c Temporal distribution of the dominant tintinnid species at Canning (S$_2$) (the annual mean and standard deviations are expressed in ind m^{-3})

Fig. 3.5 Seasonal variation of biomass and production rate at site S$_2$

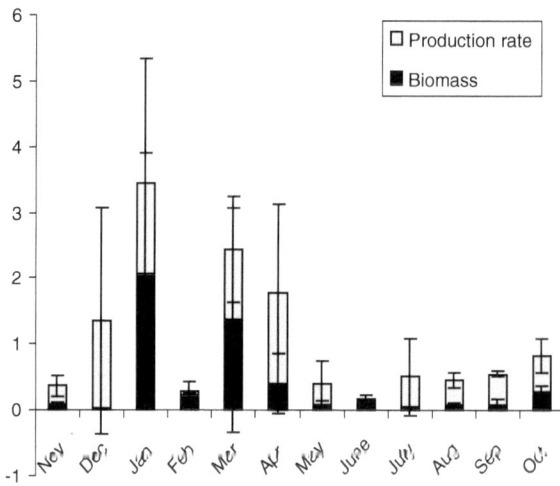

Due to dominance of the large-sized tintinnids in postmonsoon months (except *T. beroidea*), maximum biomass (4.08 µg Cl^{-1}) was recorded during January 2011 and, in contrast, minimum biomass (0.03 µg Cl^{-1}) was recorded during September 2010. The values of production rate almost coincided with the biomass accounting maximum (2.71 µg Cl^{-1} day^{-1}) during January 2011 and minimum (0.03 µg Cl^{-1} day^{-1}) during June 2011 (Fig. 3.5).

3.3.3 Lot 8 (S₃)

The maximum concentration was reached during March 2010 followed by a gradual decrease toward monsoon months. Koray and Özel (1983) found the most appropriate water temperature to be between 16 and 22 °C for most of the tintinnids in Izmir Bay, Turkey, while temperature values >24 and <10 °C stress the community. The major peak in this site coincided with a temperature of ~25 °C (Fig. 3.6).

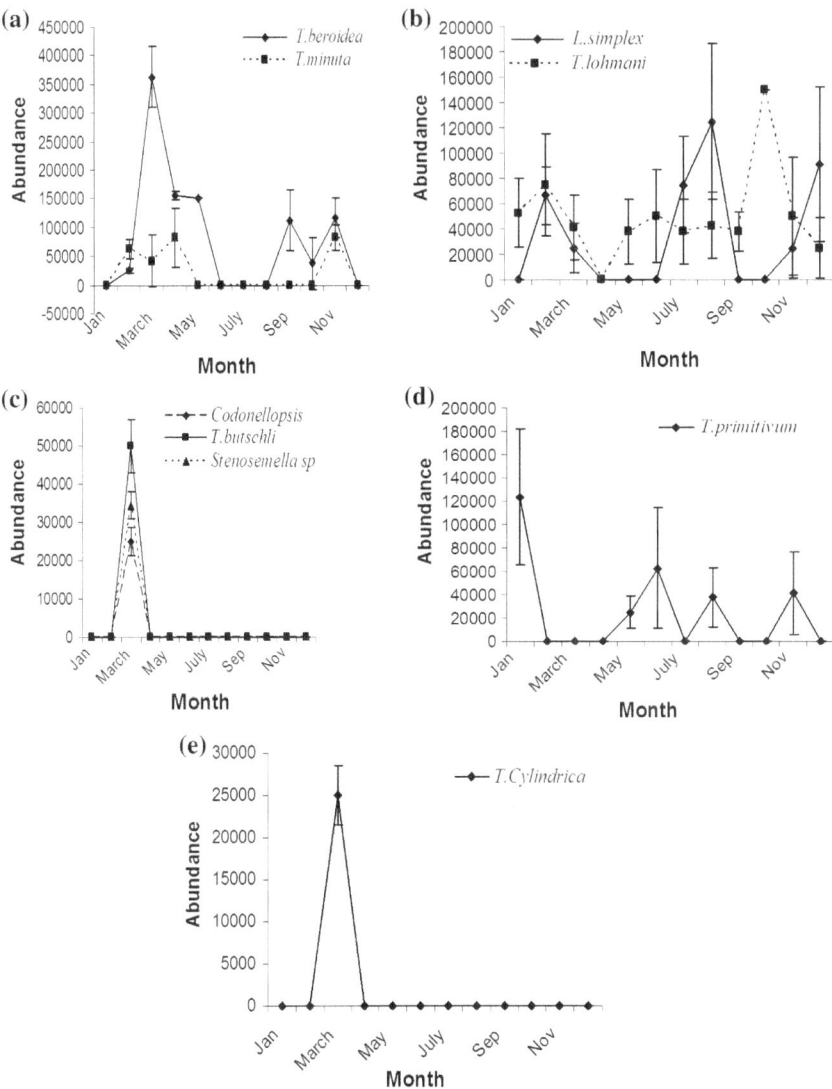

Fig. 3.6 a–e Temporal distribution of the dominant tintinnid species at Lot 8 (S₃) (the annual mean and standard deviations are expressed in ind m^{-3})

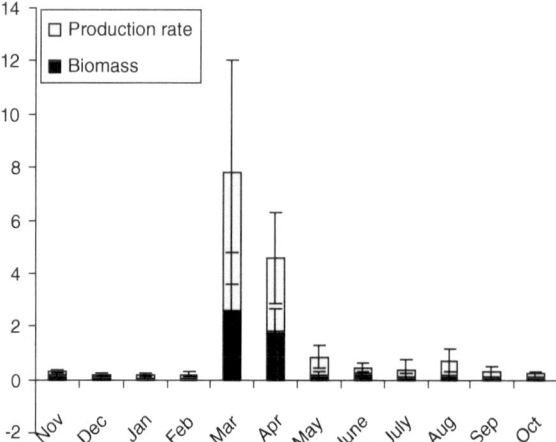

Fig. 3.7 Seasonal variation of biomass and production rate at site S_3

Both biomass and production rate coincided with the maximum abundance of tintinnid showing their maximum value (4.17 and 8.16, respectively) during March 2010 whereas their minimum values (0.06 for biomass and 0.03 for production rate) were recorded during October 11 and December 09, respectively (Fig. 3.7).

3.3.4 Chemaguri (S_4)

The site Chemaguri is a mangrove-infested site, the total abundance of tintinnid remained consistently low in this area which might be due to its high turbidity, mainly caused by intensive fishing and water transport throughout the year coupled with other anthropological stresses and subsequent low chl a concentration. The pattern of numerical abundance of tintinnids exhibited comparatively low values during premonsoon (75–150 ind l^{-1}) months with a gradual rising trend toward monsoon (82–225 ind l^{-1}) and postmonsoon (150–405 ind l^{-1}). Temporal distribution of the dominant tintinnid species has been depicted in Fig. 3.8a–d.

Biomass value coincided both with the total abundance and size of tintinnids. Maximum (1.41) and minimum (0.06) values of biomass were recorded during postmonsoon (November, 10) and Monsoon (September, 10), respectively. Maximum value (2.44) of production rate went parallel with the biomass during November 10 and the lowest (0.01) value was recorded during February 2010 (Fig. 3.9).

3.3.5 Gangasagar (S_5)

T. beroidea was also found to be the most dominant and successful species at this high-energy zone, forming the bulk of the tintinnid community (up to 60–100 %)

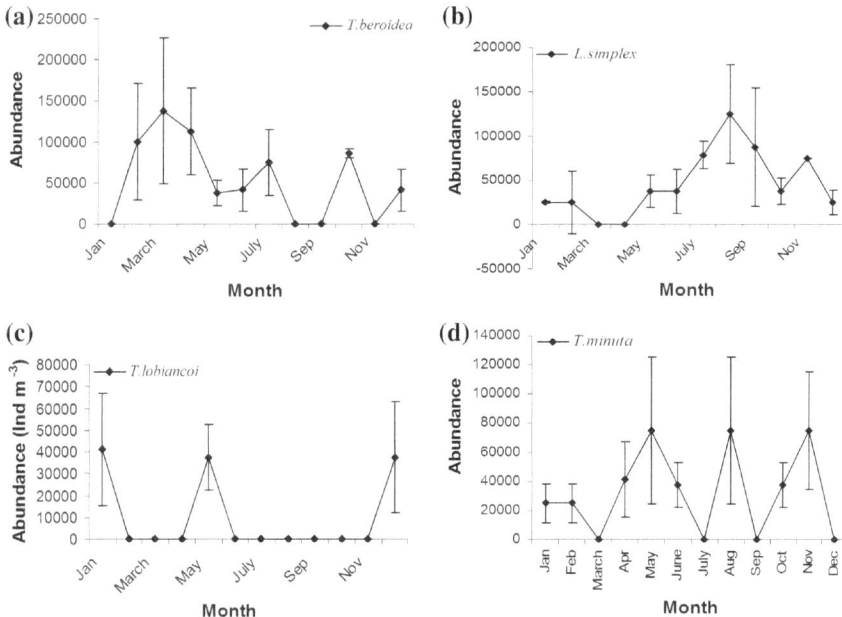

Fig. 3.8 a–d Temporal distribution of the dominant tintinnid species at Chemaguri (S$_4$) (the annual mean and standard deviations are expressed in ind m^{-3})

Fig. 3.9 Seasonal variation of biomass and production rate at site S$_4$

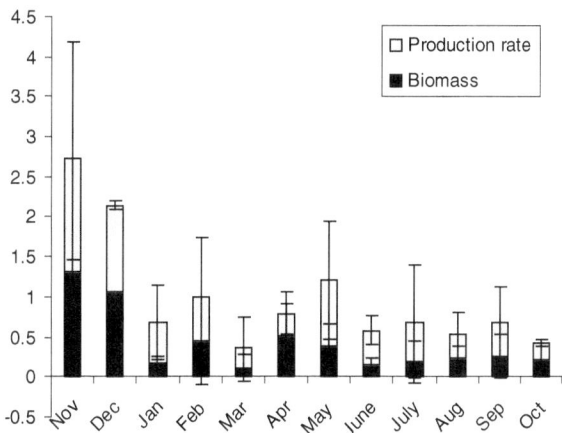

reaching maximum concentration (332 ind l^{-1}) during November, 2010. This species was followed by *T. minuta* (48–66 %) and *L. nordquisti* (24–51 %) which showed their maximum abundance during premonsoon (165 ind l^{-1}) (June, 10) and postmonsoon (312 ind l^{-1}) (January, 2011), respectively. The abundance of dominant species has been endorsed in Fig. 3.10a–c.

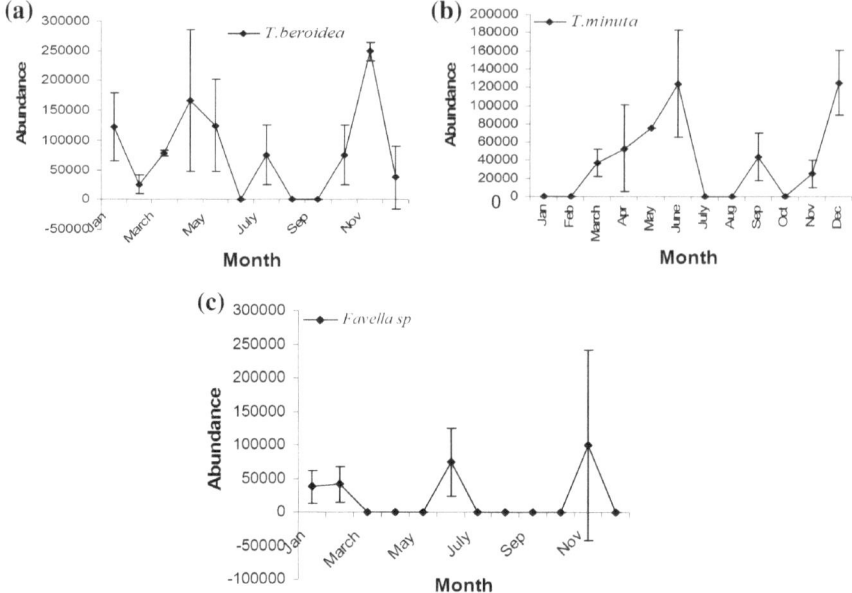

Fig. 3.10 a–c Temporal distribution of the dominant tintinnid species at Gangasagar (S₅) (the annual mean and standard deviations are expressed in ind m⁻³)

Both Biomass and production rate coincided with the total abundance and size of tintinnids. Their maximum values (16.24 μg Cl⁻¹ and 19.46 μg Cl⁻¹ day⁻¹) were documented during November 2010 coinciding with the occurrence of three large-sized tintinnids (*T. beroidea*, *T. radix* and *F. ehrenbergii*). Minimum value for biomass (0.04 μg Cl⁻¹) was recorded during June 2011 and that for production rate (0.01 μg Cl⁻¹ day⁻¹) was during February 2010 (Fig. 3.11).

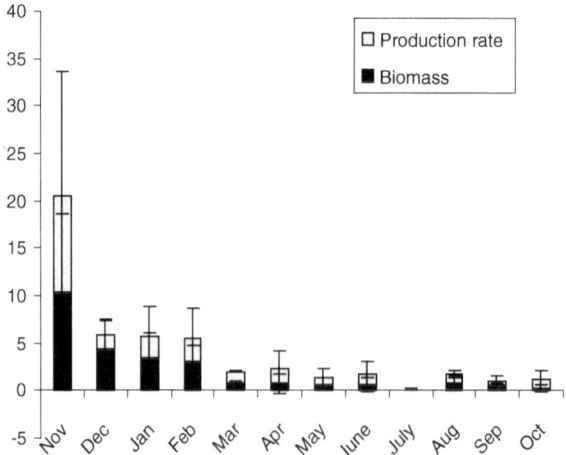

Fig. 3.11 Seasonal variation of biomass and production rate at site S₅

3.4 Tintinnids in Relation to Environmental Variables

As evident from the correlation matrix values (Table 3.2), the dominant tintinnids showed significant correlations in majority of the cases with water temperature and inorganic nutrients. Some species exhibited strong positive correlations with D.O. and B.O.D. while others are negatively correlated with turbidity and nutrient concentration. In most of the sites, tintinnid abundance was positively correlated with water temperature, salinity, and chl a. However, only few species ($L.$ $simplex$, $T.$ $cylindrica$, $T.$ $amoyensis$) showed significant correlations with nutrient (NO_3^{2-}, SiO_4^{4-}) concentrations also. Dominant species like $T.$ $beroidea$ was positively correlated with temperature ($r = 0.945$; $P \leq 0.05$), salinity ($r = 0.667$; $P \leq 0.05$), and chl a ($r = 0.668$; $P \leq 0.05$) indicating its higher abundance in open ocean. Core species like $T.$ $minuta$ showed negative correlation with nitrate ($r = -0.961$; $P \leq 0.05$) but positively correlated with phosphate ($r = 0.778$; $P \leq 0.05$) and water temperature ($r = 0.634$; $P \leq 0.05$). Again, $T.$ $lohmani$ showed positive correlation with water temperature ($r = 0.713$; $P \leq 0.05$), salinity ($r = 0.607$; $P \leq 0.05$), and dissolved oxygen ($r = 0.762$; $P \leq 0.05$). Seasonal species $T.$ $mucicola$ showed positive correlation with dissolved oxygen ($r = 0.685$; $P \leq 0.05$) and silicate ($r = 0.755$; $P \leq 0.05$) and negatively correlated with temperature ($r = 0.662$; $P \leq 0.05$). This suggests that spatiotemporal succession of species was strongly influenced by environmental variables during the study period. An overall species abundance showed a significant positive correlation with diversity ($r = 0.78$; $P \leq 0.05$) and negatively correlated with richness ($r = -0.863$; $P \leq 0.05$) indicating that existence of a particular species can lower the diversity as well as the population density in that site. Species number and abundance showed strong positive correlation for all sampling sites ($r = 0.97$; $P \leq 0.001$).

As a result, both species diversity and richness values were also very low in comparison with other sites. The synergistic effects of low salinity coupled with high turbidity values observed at this site precluded the intrusion of tintinnids. It is presumed that the presence of high concentration of small suspended particles in the water may interfere with the filter-feeding mechanism of the tintinnids (Laybourn-Parry et al. 1992). Maximum mean species diversity (2.39) and richness (9.74) were noticed in Gangasagar (S_5) and Dhamakhali (S_1), respectively (as shown in Table 3.3), whereas their minimum values were observed in Gangasagar which are related to macrotidal effects and continuous wave fluxes.

One-way ANOVA was performed for each sample to test the variation between species with months and sites taking into consideration the numerical abundance of all the species. The resulting output showed significant variation between species abundance and months ($F = 2.24$; $P = 0.004$) as well as between sites ($F = 3.3$; $P = 0.013$).

The dominance indices (Y_i) for $T.$ $beroidea$ were found to be high as 3.72 at Canning (S_2) followed by $L.$ $simplex$ (2.56) and $T.$ $primitivum$ (1.68) at Dhamakhali (S_1) indicates their abundance as well as existence throughout the year in this estuarine system. The other two less frequent ciliates $T.$ $lohmani$ (1.39) and $T.$ $minuta$ (1.21) showed their main existence in Lot 8 (S_3).

Table 3.2 Correlations between abundance of 8 dominant tintinnid ciliates and other environmental variables at five sampling sites (S_1–S_5) of Sundarbans

Species	Temp.	Salinity	pH	Turbidity	TDS	D.O.	B.O.D.	Chl a	NO_3	PO_4	SiO_4
T. beroidea	**0.945***	**0.667***	−0.509	−0.557	0.107	0.178	0.184	**0.668***	0.366	−0.299	0.446
T. minuta	**0.634***	0.074	0.164	0.495	0.552	−0.193	0.015	−0.417	**−0.961***	**0.778**	0.131
T. lohmani	**0.711***	**0.607***	0.693	−0.136	0.730	**0.762***	−0.076	0.241	−0.408	0.445	0.134
T. lobiancoi	0.190	0.036	0.384	−0.208	0.217	**−0.709***	0.593	−0.050	0.057	0.502	0.558
T. primitivum	**0.634***	−0.590	0.178	0.129	−0.517	0.107	0.601	−0.593	0.235	0.077	0.415
L. simplex	−0.456	−0.372	−0.593	−0.553	−0.444	−0.043	−0.098	0.057	**0.699***	−0.682	0.243
T. mucicola	**−0.662***	−0.175	0.254	−0.285	0.135	**0.685***	0.440	−0.271	0.160	0.330	**0.755***

Bold and asterix values denotes significance at 5 % level (P<0.05)

Table 3.3 Community index values of tintinnids for five sampling sites (S$_1$–S$_5$)

	Dhamakhali S$_1$	Canning S$_2$	Lot 8 S$_3$	Chemaguri S$_4$	Gangasagar S$_5$
Species diversity	1.2	1.83	1.76	1.97	2.34
Species richness	9.74	8.65	8.88	8.07	7.38
Species evenness	0.36	0.25	0.45	0.41	0.54

A dendrogram of eight dominant tintinnids encountered from five sampling sites was plotted using group-mean clustering from log-transformed data (Fig. 3.12). A single cluster was formed by five postmonsoon species (*T. lohmani, T. primitivum, T. lobiancoi, T. tubulosa,* and *T. mucicola*) followed by three outliers among which the first one (*T. minuta*) is a typical premonsoon species and the other two (*T. beroidea* and *L. simplex*) are available almost throughout the year with maximum density in premonsoon.

K-dominance curves were plotted to allow a better comparison of differences in tintinnid diversity between the investigated stations (Fig. 3.13). The most elevated curve showed the lowest diversity at the site Chemaguri (S$_4$). The dominance of tintinnid species between stations was found to be similar in reference to species rank. A dominance >80 % was found only at S$_4$ at species rank 8, while at stations S$_2$ and S$_5$, dominance >80 % was reached at species rank 9 and stations S$_1$ and S$_3$ showed >80 % dominance at species rank 10 or more.

To gain insight into the spatial coincidences in the distribution of tintinnids in relation to environmental variables (temperature, salinity, pH, turbidity, chl *a*, dissolved oxygen, biochemical oxygen demand, nitrate, phosphate, and silicate

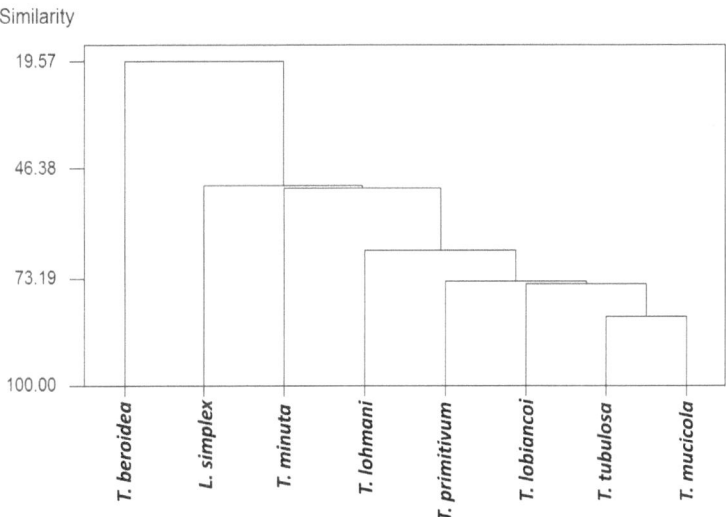

Fig. 3.12 Dendrogram showing the season wise clustering of 8 dominant tintinnids found in Sundarbans

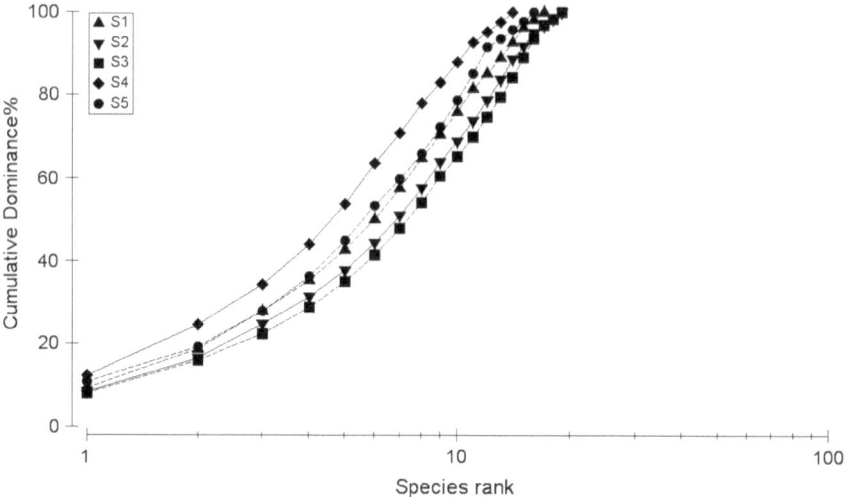

Fig. 3.13 K-dominance curves of tintinnid species (*x* axis logged) for five stations (S$_1$–S$_5$) in Sundarbans coastal waters, where most *elevated curve* showed lowest diversity

concentration in surface water) involved, a principal component analysis (PCA) was performed, which revealed three main groups (Fig. 3.14). Group I represents some frequently available species (*T. directa, T. lobiancoi, T. kofoidi, L. simplex, T. tubulosa,* and *S. ventricosa*) attributed mainly by turbidity, pH, B.O.D., and Phosphate concentration. These species appeared in low numbers, being more frequent in premonsoon and postmonsoon mainly. The second group was formed by *T. beroidea, T. minuta, L. nordquisti, T. lohmani, T. cylindrica, T. urnula, F. ehrenbergii, T. primitivum, T. gracilis,* and *T. mucicola,* probably regulated by chl *a*, salinity, and temperature. This group was mainly formed by small-sized taxa, which reached some of the greatest abundances of the estuary. Group III composed of nine tintinnid species (*T. amoyensis, T. radix, T. parva, T. butschlii, T. mortensenii, C. schabi, T. uruguayensis, T. parvula,* and *Helicostomella* sp.), which were present only once throughout the study, mainly appearing at the seaward end of the estuary. Species in this group were mainly influenced by environmental variables such as nitrate, silicate, and D.O.

Multiple stepwise regression model between abundance of eight dominant tintinnid and environmental factors revealed that nitrate was found to be positively related with majority of cases (/) with density of total tintinnid. The next factor was chl *a* and temperature, both found to be positively related in six cases. The best multiple regression models were produced by a core species *T. beroidea* ($R^2 = 83.9$ %; $P = 0.004$) and two premonsoon species *T. minuta* ($R^2 = 80.6$ %; $P = 0.001$) and *L. simplex* ($R^2 = 87.6$ %; $P = 0.000$) as those species were best fitted with the environment in that specific season. Multiple regression models with other species were not significant suggesting that the species can withstand in fluctuating environmental conditions.

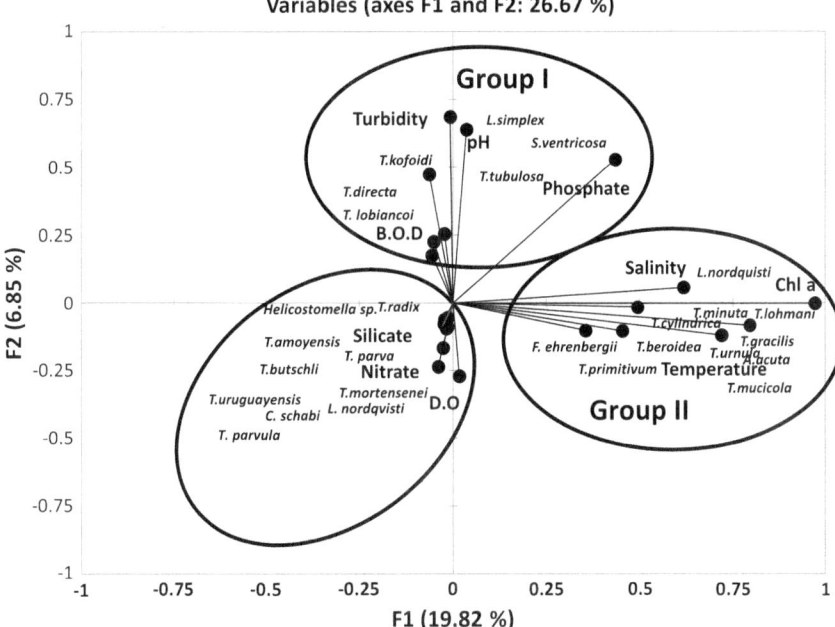

Fig. 3.14 Principal component analysis (PCA) map of tintinnid species in relation to environmental variables

3.5 Polymorphism in Tintinnids

Polymorphism in biology occurs when two or more clearly different phenotypes exist in the same population of a specie, in other words, the occurrence of more than one *form* or *morph*. In order to be classified as such, morphs must occupy the same habitat at the same time and belong to a panmictic population (one with random mating) (Ford 1965). Polymorphism results from evolutionary processes, as does any aspect of a species. It is heritable and is modified by natural selection. Polymorphism is common in nature; it is related to biodiversity, genetic variation, and adaptation; it usually functions to retain variety of forms in a population living in a varied environment (Dobzhansky 1970).

Few species of tintinnid showed a pattern of polymorphism in nature. Seasonal patterns in the oral diameter and length of the lorica of *T. beroidea* reflected decrease in the size of the individual species with decreasing temperature (Fig. 3.15). The smallest lorica (17–25 μm in length and 4.76 μm in lorica oral diameter) was recorded in postmonsoon (January) period when the temperature was ~18 °C. In premonsoon with a temperature of ~32 °C, the length of the lorica ranged between 80 and 110 μm lorica oral diameters ranged between 16 and 18.5 μm.

This difference between premonsoon and postmonsoon specimens could be explained in the context of physiological processes which might be accelerated as a

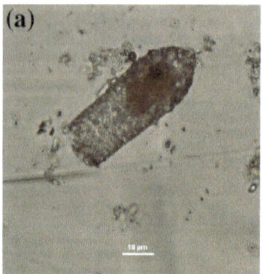

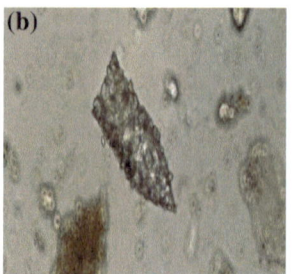

Length= 95.4 µm   **Length=50.6 µm**
Lorica oral diameter=17.2 µm   **Lorica oral diameter=10.5 µm**

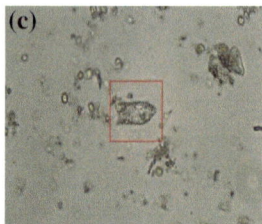

Length= 17.25 µm
Lorica oral diameter=4.76 µm

Fig. 3.15 Polymorphism of *T. beroidea* at different season (**a**, **b**, **c**) of the year at the site Canning (S_2). **a** Premonsoon. **b** Monsoon. **c** Postmonsoon

function of elevated temperature (Gold and Morales 1975). Low salinity during early postmonsoon may also be responsible for the small size of *T. beroidea*. Other important parameters viz. chl *a* and nutrient concentrations were sufficiently high during postmonsoon season and thus cannot be the reason for the smaller size of tintinnids. *L. simplex* also showed reduction in body size during unfavorable environments.

3.6 Comparative Account of Tintinnid in Coastal Regions

3.6.1 Parangipettai, Southeast Coast of India

The distribution and abundance of tintinnids varied remarkably due to the seasonal environmental fluctuations in the Vellar estuarine and mangrove waters, total abundance of tintinnids was maximum in summer and postmonsoon than the other seasons, as commonly observed in many marine coastal and estuarine waters. The abundance range was recorded 2–420 ind l^{-1}. The higher proliferation of tintinnids in summer months might be attributed to higher temperature and chl *a* concentrations, both are often considered as the most important factors to increase the growth rate.

The abundance of tintinnids was lowest during monsoon months, when the water column was markedly stratified to a large extent because of heavy rainfall, overcast sky, and cool conditions. As a result of these, water temperature, salinity, and chl *a* concentration decreased largely with increased turbidity. The low temperature and chl *a* concentrations drastically affect the life cycles of many tintinnids. Hence, the influence of these factors severely affects the abundance and growth rate of tintinnids. Total of 47 species belonging to 14 genera of tintinnids was identified in this estuarine and mangrove waters; of these, 20 species belonged to the genus *Tintinnopsis*. The community was dominated by a cluster of premonsoon species (*Tintinnopsis minuta, T. Beroidea, T. Mortensenii, T. Uruguayensis, Codonellopsis ostenfeldii, Favella philippinensis, Tintinnopsis lohmani, Tintinnopsis kofoidi, T. bermudensis, T. tenus, Stensomella nivalis, Coxliella annulata, Helicostomella longa, Rhabdonella spiralis, Dadayiella ganymedes, Tintinnidium incertum, T. primitivum;* monsoon species (*Tintinnopsis cylindrica, Eutintinnus tenus, L. simplex*) and postmonsoon species (*T. incertum, Tintinnopsis butschlii, T. gracilis, T. glans, T. nucula, T. radix, T. tubulosa, F. ehrenbergii, S. ventricosa, Codonellopsis schabi, Dictyosysta seshaiyai,* and *F. brevis*). They are often common and contributed significantly to the total abundance of tintinnids, as reported in many coastal and oceanic waters indicating that their adaptability to different thermal and salinity gradients.

3.6.2 Bay of Bengal and Andaman Sea

The diversity of tintinnids was studied in the Bay of Bengal and Andaman Sea during inter-monsoon (March) period. Tintinnids comprised of 36 species assigned to 18 genera and were mainly represented by the following species: *Eutintinnus fraknoi, Undella claperedei, R. spiralis, Salpingella acuminata, Protorhabdonella simplex, Salpingacantha ampla, Amphorella pyramidata, Steenstrupiella steenstrupi, Epiplocylis undella, Petalotricha ampulla,* and *Dictyocysta mitra*. The tintinnids ciliates comprised 19 % among all the microzooplankton community. Microzooplankton density and biomass varied from 6 to 115×10^3 ind l^{-1} (mean 45.7×10^3 ind l^{-1}) and 0.17 to 2.7 mg cm^{-3} (mean 1.1 mg cm^{-3}), respectively. The maximum numerical abundance and biomass were found in the mixed layer where chl *a* reaching maximum value. The density was decreased gradually with increasing depth. Some species were restricted within 75–100 m of depth (*Salpingella* and *Salpingacantha*). Preference to particular environmental factors like temperature and salinity might have influenced their vertical distribution. A significant positive correlation was obtained between chl *a* and microzooplankton biomass.

3.6.3 Cochin Backwaters

In Cochin backwaters (CBW), a tropical estuarine system located along the southwest coast of India, microzooplankton was represented by 36 ciliate species in which tintinnid was dominated comprising of 22 species. The abundance of microzooplankton was markedly higher during presummer monsoon (mean $3,817 \pm 1,893$ ind l^{-1}) compared with the onset of summer monsoon (mean $2,050 \pm 1,623$ ind l^{-1}) and peak summer monsoon (mean 409 ± 184 ind l^{-1}). Salinity is reported to play a major role in controlling the distribution of tintinnids in estuaries (Godhantaraman 1994; Sujatha and Panigrahy 1999). Stations with high salinity always formed a single cluster indicating that the microzooplankton abundance in the system is controlled by salinity. Distinctly low species diversity, richness, and evenness during summer monsoon suggest that only a few species are tolerant to near zero values of salinity. Eight species of ciliates viz., *Tintinnopsis lohmani, T. beroidea, T. tubulosa, T. uruguayensis, T. incertum, L. simplex, Euplotes vannus,* and *Strombidinopsis cheshirii* were present during peak summer monsoon exhibiting wide range of tolerance to salinity. The environmental variables such as temperature and dissolved oxygen did not show any significant correlation with microzooplankton abundance suggesting that these parameters do not control the abundance and distribution of microzooplankton in this system. Loricate ciliate was present during all the surveys, accounting maximum abundance during the onset of summer monsoon (82 %) dominated by *Tintinnopsis.*

3.6.4 Central and Eastern Arabian Sea

The tintinnid species recorded in the Central and Eastern Arabian Sea in all the three seasons of monsoon (Inter-monsoon, Winter monsoon, and Summer monsoon) were represented by *Tintinnopsis* sp., *Eutintinnus* sp., *Favella* sp., *Parundella* sp., *Amphorides* sp., *Codonellopsis* sp., *Rhabdonellopsis* sp., *Dictyocysta* sp., *Codonella* sp., *Tintinnus* sp., *Xystonellopsis* sp. Other genera that are found were comprised of *Tintinnidium* sp., *Metacylis* sp., *Ascambelliella* sp., *Parafavella* sp., *Coxliella* sp., *Dadayiella* sp., *Protorhabdonella* sp., *Rhabdonella* sp., *Salpingella* sp., *Stensomella* sp., *Helicostomella* sp., *Leprotintinnus* sp., and *Undella* sp., etc. The microzooplankton abundance during three seasons in the Central and Eastern Arabian Sea exhibited mean cell concentration of 500 ind l^{-1}. The abundance of tintinnids was maximum during inter-monsoon (700 l^{-1}) followed by summer (300 ind l^{-1}) and winter season (130 ind l^{-1}). Higher abundance of microzooplankton was found in the upper 100 m compared to 100–200 m depth during inter-monsoon period. The density of microzooplankton was higher in open ocean water during inter-monsoon and winter, while it was more at coastal stations during summer. Maximum population of microzooplankton occurred during inter-monsoon when primary production and chl *a* concentration were low but bacterial population was high. Thus, it is assumed that microzooplankton population increased through a microbial loop during this season. The higher population of microzooplankton in coastal waters during summer monsoon might have resulted due to river runoff.

References

Abboud-Abi Saab, M. (1985). Étude quantitative et qualitative du phytoplancton des eaux côtières libanaises. *Lebanese Science Bulletin, 1*, 197–222.

Abboud-Abi Saab, M. (1989). Distribution and ecology of tintinnids in the plankton of Lebanese coastal waters (eastern Mediterranean). *Journal of Plankton Research, 2*, 203–222.

Abhay Kumar, V. K., & Dube, H. C. (1995). Occurrence and distribution of bacterial indicators of fecal pollution in the tidal water of a muddy coast. *Journal of the Marine Biological Association of India, 37*, 98–101.

Aleya, L. (1991). The concept of ecological succession applied to a eutrophic lake through the seasonal coupling of diversity index and several parameters. *Archiv für Hydrobiologie, 120*, 327–343.

Barría de Cao, M. S., Pettigrosso, R. E., & Popovich, C. (1997). Planktonic ciliates during a diatom bloom in Bahía Blanca Estuary, Argentina. II. Tintinnids. *Obelia, 23*, 21–31.

Capriulo, G. M., & Carpenter, E. J. (1983). Abundance, species composition and feeding impact of tintinnid micro-zooplankton in central Long Island Sound. *Marine Ecology Progress Series, 10*, 277–288.

Dobzhansky, T. (1970). *Genetics of the evolutionary process*. New York: Columbia University Press.

Ford, E. B. (1965). *Genetic polymorphism*. London: Faber & Faber. ISBN 9780262060127.

Godhantaraman, N. (1994). Species composition and abundance of tintinnids and copepods in the Pichavaram mangroves (South India). *Ciencias Marinas, 20*, 371–391.

Godhantaraman, N. (2002). Seasonal variations in species composition, abundance, biomass and estimated production rates of tintinnids at tropical estuarine and mangrove waters, Parangipettai, southeast coast of India. *Indian Journal of Marine Sciences, 36*, 161–171.

Godhantaraman, N., & Uye, S. (2001). Geographical variations in abundance, biomass and trophodynamic role of microzooplankton across and inshore-offshore gradient in the Inland Sea of Japan and adjacent Pacific Ocean. *Plankton Biology and Ecology, 48*, 19–27.

Gold, K., & Morales, E. A. (1975). Seasonal changes in lorica sizes and the species of Tintinnida in New York Bight. *Journal of Protozoology, 22*, 520–528.

Kannan, R., & Kannan, L. (1996). Physicochemical characteristics of seaweed beds of the Palk Bay, southeast coast of India. *Indian Journal of Marine Science, 25*, 358–362.

Kimor, B. (1969). The occurrence of some tintinnid protozoans in the plankton of Lake Tiberias. *Verhandlungen—Internationale Vereinigung fuer Theoretische und Angewandte Limnologie, 17*, 358–361.

Kimor, B., & Golandsky, B. (1977). Microplankton of the Gulf of Elat: Aspects of seasonal and bathymetric distribution. *Marine Biology, 42*, 55–67.

Koray, T. & Özel, I. (1983). Izmir Korfezi planktonundan saptanan *Tintinnoinea turleri*. I. Ulusal Deniz ve Tatlisu Arastirmalari Kongresi Tebligleri. 15–17 Ekim 1981, *Izmir. E.U.F.F. Derisi, Series B I*, 221–244.

Laybourn-Parry, J., Marchant, H. J., & Brown, P. E. (1992). Seasonal cycle of the microbial plankton in Crooked Lake. *Antarctica, Polar Biology, 12*, 411–416.

Lebour, M. V. (1922). The food of plankton organisms. *Journal of the Marine Biological Association of the United Kingdom, 12*, 644–647.

Mahajan, A., & Kanhere, R. L. (1995). Seasonal dynamics of phytoplankton population in relation to abiotic factors of a fresh water pond at Barwar (MP). *Pollution Research, 14*(3), 347–350.

Marichamy, R., Gopinathan, C. P., & Siraimeetan, P. (1985). Studies on primary and secondary production in relation to hydrography in the inshore waters of Tuticorin. *Journal of Marine Biology Association of India, 27*, 129–137.

Menon, N. R., Venugopal, P., & Goswami, S. C. (1971). Total biomass and faunistic composition of zooplankton in the Cochin backwaters. *Journal of Marine Biology Association of India, 12*, 220–225.

Modigh, M., & Castaldo, M. (2002). Variability and persistence in tintinnid assemblages at a Mediterranean coastal site. *Aquatic Microbial Ecology, 28,* 299–311.

Mukhopadhyay, S. K., Biswas, H., De, T. K., & Jana, T. K. (2006). Fluxes of nutrients from the tropical River Hooghly at the land–Ocean boundary of Sundarbans, NE coast of Bay of Bengal India. *Journal of Marine Systems, 62,* 9–21.

Nielsen, T. G., & Kiorboe, T. (1994). Regulation of zooplankton biomass and production in a temperate, coastal ecosystem. 2 Ciliates. *Limnology and Oceanography, 39,* 508–519.

Reynolds, C. S., (1997). Vegetation processes in the pelagic: A model for ecosystem theory. In *Excellence in ecology* (Vol 9, 371 pp). Oldendorf: Ecology Institute.

Reynolds, C. S. (2006). *The ecology of phytoplankton.* Cambridge: Cambridge University Press. 535 pp.

Sujatha, M., & Panigrahy, R. C. (1999). The tintinnids (Protozoa: Ciliata) of Bahuda estuary, east coast of India. *Indian Journal of Marine Sciences, 28,* 219–221.

Urrutxurtu, I., Orive, E., & Sota, A. (2003). Seasonal dynamics of ciliated protozoa and their potential food in a eutrophic estuary (Bay of Biscay). *Estuarine, Coastal and Shelf Science, 57,* 1169–1182.

Vincent, D., & Hartmann, H. J. (2001). Contribution of ciliated microprotozoans and dinoflagellates to the diet of three species in the Bay of Biscay. *Hydrobiologia, 443,* 193–204.

Chapter 4
Impact of Stress on Tintinnid Community

Abstract Microzooplankton play a number of pivotal ecological roles at the base of marine food webs, and it is very important to understand the effects of stresses on the species composition, biomass, and trophic activities. The chapter deals with the tremendous negative impact on water quality as well as plankton community structure due to occurrence of profuse algal bloom and tropical cyclonic storm "Aila" and thus necessitates sound management strategies. The author recommends for innovative research on the phonological changes in plankton due to climate change, invasion of alien species by modern technologies, and novel application of ecosystem-based adaptation (EBA) for protection and restoration of Sundarbans wetland ecosystem.

Keywords Algal bloom · Invasive species · Community structure · Water quality · Climate change · Natural geohazards

4.1 Effect of Algal Bloom by Centric Diatom *Hemidiscus hardmannianus* (Bacillariophyceae)

Algal bloom, a natural phenomenon resulting from a coupling mechanism involving physical, chemical, and biological factors, is one of the important biological nuisances to the coastal region. Incidence of algal blooms and discolorations of water in the Indian seas have been reported by Devassy and Nair (1987). Studies on algal blooms using the remote sensing satellite data along the west coast of India have been worked out by Sarangi et al. (2005).

Algal blooms are the subject of major global interest during the last two decades. Phytoplankton has 3,400–4,100 microalgae species, and among them, 300 species could produce blooms (Smayda 1997; Naz et al. 2012). Accordingly, different species might be involved in algal blooms in different regions (Hu et al. 2011). Phytoplankton taxon abundance depends upon the growth, immigration, physical concentration, and other mechanisms based upon the physical, chemical,

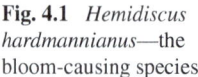
Fig. 4.1 *Hemidiscus hardmannianus*—the bloom-causing species

and biological characteristics of water and sediment fluctuation. In this regard, many species generally struggle to coexist in a water mass of seemingly similar properties. Hence, one species becomes numerically dominant than the other forms and occurs as a monospecific bloom which is a remarkable phenomenon, especially dependent upon the nature of environment (Sachithanandam et al. 2013). Implications from phytoplankton blooms may result in adverse effects on aquatic species and their ecosystems (Monaldo et al. 1997). One significant impact includes the blockage of incoming solar radiation, which results in a disturbance of the aquatic habitat (Diersing 2009). Phytoplankton blooms play a key role in pelagic and benthic food webs, on the transfer of organic matter to the seafloor as well as on ocean biogeochemistry (Falkowski et al. 1998).

On July 2010, a dense bloom of *Hemidiscus hardmannianus* [(Greville) Mann 1907] (Bacillariophyceae) (Fig. 4.1) was noticed at three sites of Sundarbans mangrove environments, namely Lot 8 (S_3), Chemaguri (S_4) and Gangasagar (S_5) covering the western part of Sundarbans (as shown in Fig. 2.1). The species is open marine, planktonic, and most common in warm waters (Round et al. 1990). The bloom was extended to a surface area of ~3–4 km, resulting a greenish color along with an obnoxious smell from the surface water and persisted for almost one week.

4.1.1 Changes in Water Quality

The pronounced changes in water quality characteristics have been shown in Table 4.1, caused by the algal bloom. The water temperature was comparatively high with a range from 27.5 to 33.2 °C, which was already been recognized as an important factor that induces the bloom condition of *H. hardmannianus* (Round et al. 1990). The observed salinity ranged from 7.1 to 29.8 p.s.u. A gradual increase in salinity was noticed from the upper part of the estuary (Lot 8; S_3)

Table 4.1 Water quality parameters during prebloom, bloom, and postbloom conditions from three sampling sites of Sundarbans wetland

Parameters	Lot 8 (S$_3$)			Chemagari (S$_4$)			Gangasagar (S$_5$)		
	Prebloom	Bloom	Postbloom	Prebloom	Bloom	Postbloom	Prebloom	Bloom	Postbloom
Water temperature (°C)	33	27.5	30	32	28.3	33	33	28	32
Salinity (p.s.u.)	15	7	6.9	21	11.3	14	30	29.8	15
pH	8.13	8.6	8.3	8.1	8.7	8.1	7.9	8.6	7.9
Turbidity (NTU)	16	2	12	18	2	18	8	1	11
D.O. (mg/L)	4.48	5.68	3.7	4.35	5.19	3.41	4.22	5.09	3.9
B.O.D. (mg/l)	1.62	1.16	2.67	1.71	1.41	1.38	2.12	0.64	1.71
Nitrate (μmol l^{-1})	22.11	3.71	20.74	15.14	1.07	22.68	7.2	1	12.17
Phosphate (μmol l^{-1})	1	0.11	2.08	0.85	0.08	2.95	0.33	0.02	2.5
Silicate (μmol l^{-1})	79.62	46.11	90.98	58.18	18.02	65.34	41.56	25.93	61.64
Chl a (mg/m^3)	1.77	8.53	1.5	2.78	8.8	1.19	1.7	8.59	0.8

toward the open ocean (Gangasagar; S_5). The pH value increased a few fold (8.1–8.6) during bloom period than pre- and postbloom phase (mean 8.2). This might be due to consumption of CO_2 in water (HCO_3^+) by large amount of phytoplankton mass during photosynthesis and thus reduction of H^+ ion and increase of alkalinity which ultimately results in high pH.

The minimum and maximum values of dissolved oxygen (D.O.) were observed during postbloom (3.51 mg l^{-1}) and bloom phases (5.68 mg l^{-1}). Relatively, high D.O. concentration noticed during bloom phase might be due to photosynthetic release of O_2 by the dense diatom mass which was also endorsed by previous workers (Pielou et al. 1969). As expected, very low D.O. contents were observed during postbloom period indicating the whole phytoplankton mass is in decayed stage and the bacteria responsible for the decay process are consuming most of the D.O. in water. In association with low D.O. concentration (~3.67 ± 0.24 mg l^{-1}), the B.O.D value was significantly high (~1.92 ± 0.67 mg l^{-1}) during postbloom and relatively low in bloom condition.

During the peak bloom period, siltation rate in the surface water was very low resulting unusual lower values of turbidity (1.6 ± 0.29 NTU) which was previously reported by Subramanian and Purushothaman (1985). On the other hand, turbidity was extremely high during postbloom phase (13.66 ± 2.57 NTU), when the entire diatom sedimented at the bottom as dead mass.

The observed values of the inorganic nutrients showed wide range of variations in prebloom, bloom, and postbloom phases (Table 4.1). The concentration ranges of nitrate, phosphate, and silicate are 2.68–3.71 μmol l^{-1}, 0.11–2.95 μmol l^{-1}, and 18.93–90.98 μmol l^{-1}, respectively (expressed in mean values). Interestingly, three nutrients showed maximum and minimum values during postbloom and bloom phases, respectively. Apart from physical and chemical processes, phosphate concentration in coastal waters mainly depends upon phytoplankton uptake and replenishment by microbial decomposition of organic matter (Pielou et al. 1969). It is reported that sinking of decomposing plankton in seawater results in oxygen consumption and phosphate liberation. Also, bacteria liberate phosphate from dead phytoplankton added to seawater (Satpathy et al. 2007). The high phosphate concentration in postbloom phase may be attributed to such reasons. Silicate is utilized for the formation of the siliceous frustules of diatoms and constitutes one of the most important nutrients regulating the phytoplankton growth and proliferation and ultimately blooming. Hence, it is evident that an overall reduction in nutrients was marked during the bloom period which supported a rise in phytoplankton productivity in the area in relation to the pre- and postbloom phases (Jayaraman et al. 1937).

Concentrations of chlorophyll a (chl a) also showed wide range of variations (0.8–8.9 mg m^{-3}). This showed an exponential increase (8.64 mg/m^3) coinciding with the highest cell density (8.86 × 10^6 cells l^{-1}). In general, concentration of chlorophyll remained high during bloom as compared to pre- and postbloom period, and the peak value of the pigment was about 8–9 times higher in comparison to the non-bloom period. Similar observations of unusually high pigment concentration have been reported by Satpathy et al. (2007) during bloom phases

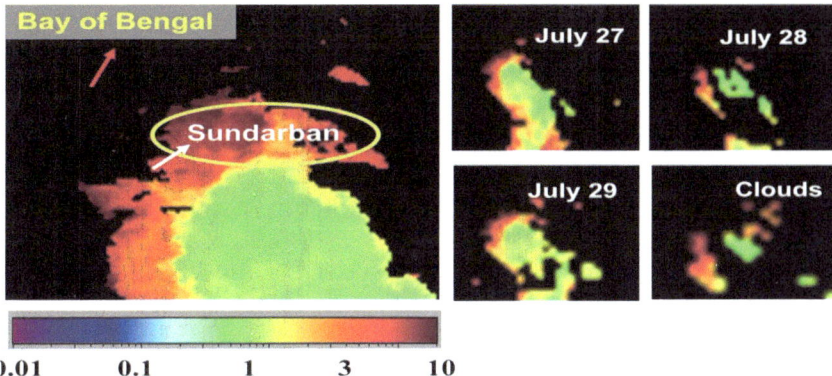

Fig. 4.2 MODIS-Aqua satellite-derived chlorophyll maps shown over the Bay of Bengal water (2010)

from eastern coastal part of India. Satellite remote sensing-derived chlorophyll images have been retrieved from the MODIS-Aqua sensor data which also endorsed enriched concentration of chlorophyll (4.0–8.0 mg m^{-3}) in this coastal region.

4.1.2 Satellite-Derived Sea Surface Temperature (SST) and Wind Speed Information

TMI satellite-derived sea surface temperature (SST) and wind speed maps have been interpreted. The SST observed to be warmer (28–32 °C) and wind speed in the range of 5–7 m/s. Hence, it is depicted that the diatom species *H. hardmannianus* blooms in higher temperature and moderate wind speed condition (Fig. 4.2).

4.1.3 Changes in Phytoplankton Community Structure

Phytoplankton community showed distinct variations for both qualitative and quantitative aspects during bloom phase. Total 61 species of phytoplankton comprising 55 diatoms, 4 dinoflagellates, and 2 cyanobacteria was identified. An overwhelming dominance of the bloom-causing diatom *H. hardmannianus* was encountered contributing ~100 % of the total phytoplankton. The species appeared as a sudden spurt in cell number when the phytoplankton population density increased abruptly from 12.84×10^3 cells l^{-1} to 8.86×10^6 cells l^{-1} resulting more than threefold increases during bloom. The observed density of *Hemidiscus* was found to be significantly higher than the earlier reported value of 49×10^3 cells l^{-1} by Subramanian and Purushothaman (1985) in Parangipettai coast, southern part of India, and by Shetty and Saha (1971) in coastal waters of Sundarbans wetland.

Saunders and Glenn (1969) considered a species to be a bloom-former when its density at any one time would produce a total surface area of 10^7 μm^2 1^{-1}. Accordingly, Santhanam et al. (1976) computed the surface area of a single cell of *H. hardmannianus* as 117,068 μm^2 and pointed out that 85 cells 1^{-1} were required to comprise a bloom. In the present study, the surface area has been measured as 187,043 μm^2, and as per above reference, only 54 cells are required to form a bloom which persisted for a week period and could be categorized as a periodic event (Saunders and Glenn 1969). The huge numerical density and large surface area of this diatom reported here certainly formed a bloom.

4.1.4 Changes in Tintinnid Community Structure

We have considered exclusively the ciliate loricate tintinnids as there are distinct advantages in examining diversity trends among tintinnids compared to other group of microzooplankters. While numerically a minority component, they are nonetheless much more abundant than foraminifera or radiolarians (Santharam et al. 1976) and there is a wealth of data on their ecology (Takahashi et al. 1977). Thus, tintinnids are a group united ecologically as microzooplankters, morphologically as loricate ciliates, and phylogenetically as members of the order Tintinnida (Dolan et al. 2002).

During the bloom period, total absence of tintinnid was registered which might be associated with high-density pressure of *H. hardmannianus* which almost outnumbered rest of the plankters. An overall dominance of *Tintinnopsis minuta* was recorded during prebloom period for all the three sites. The values of species richness and diversity indexes were as follows: 6.15 and 0.70 in Lot 8 (S₃), 2.19 and 0.69 in Chemaguri (S₄), and 1.72 and 0.70 in site Gangasagar (S₅), respectively. All the values drastically dropped down to zero during bloom period when the numerical abundance of tintinnid became absolutely zero. However, there were dramatic changes during postbloom period when *T. minuta* emerged as dominant species at Chemaguri (S₄) as before and the species richness and diversity index again raised to 2.16 and 0.70, respectively, whereas *Tintinnopsis directa* and *Undella* sp., became the leading species at Station 3 (Lot 8) and Station 5 (Gangasagar), respectively. Interestingly, the *Undella* sp., the oldest representative of tintinnids (Li et al. 2009), occurred exclusively at this period which is a deep-sea dweller in nature. The average biomass and production rate of tintinnid in the prebloom period were 0.09 μg Cl^{-1} and 0.32 μg Cl^{-1} day^{-1}, respectively, which became zero during the bloom period due to total absence of tintinnid. Later on, this regained the previous values (0.08 μg Cl^{-1} and 0.30 μg Cl^{-1} day^{-1}) during postbloom period.

The similar phenomenon was also reported by Barría de Cao et al. (1997) from Bahía Blanca estuary, Argentina, during occurrence of an algal bloom.

The observed poor density of tintinnids might be related to lack of adequate food (Barría de Cao et al. 2005) which might be related to cell size phytoplankton prey (Barría de Cao et al. 1997). It is noteworthy to mention that few species of tintinnids (only those that have large peristome and oral lorica diameters) are able to graze on phytoplankton during the bloom (Barría de Cao et al. 2005). Here, the cell size of the bloom-forming species (*H. hardmannianus*) is sufficiently large (117,068 μm^2 in diameter) making it difficult to be ingested by the tintinnids and indicating that they are not suitable food for any tintinnids in general (Barría de Cao et al. 2005). The tintinnid community regained its previous phase when this specific diatom almost disappeared allowing the normal phytoplankton community structure to be established. However, a contrasting feature was also reported from Great South Bay, New York, by Lynn et al. (1991) during a brown tide outbreak, and it is interpreted that the food condition was favorable for these tiny organisms.

4.1.5   Changes in Mesozooplankton Community Structure

Mesozooplankton composition was very unusual during the blooming period as compared with the pre- and postblooming period. The greatest impoverishment of mesozooplankton was noticed during bloom period like tintinnids when very low density of nauplius, zoea, and *Lucifer* sp., (52.91, 62.91 and 15.01 ind m^{-3} respectively) were recorded. In contrast, during prebloom period, the average numerical abundance of mesozooplankton was 1,086 ind m^{-3} in which copepod constituted the major group accounting ~76.94 % of the total zooplankton.

Copepods were dominated during pre- and postbloom periods at all three sites, especially the calanoids (contributing ~59.71–97.15 % of total mesozooplankton). Predominance of calanoid copepods in marine and estuarine waters under normal conditions is well established worldwide including Sundarbans coastal waters Duguay et al. (1989). However, an abrupt contrast feature was marked during the bloom period when copepods were either absent or represented by few species. The epipelagic chaetognath, *Sagitta bedoti*, was also absent during the bloom condition. For mesoplankters, polychaete larvae appeared in considerable numbers during the bloom period. Decapod larvae (*Penaeus monodon, P. indicus, Metapenaeus monoceros, M. dobsoni,* and *M. affinis*) were found in considerable numbers only during pre- and postbloom conditions.

4.1.6   Statistical Interpretations

The diversity index indicates the degree of complexity of a community structure (Sarkar et al. 2007). The diversity index (H') varied between stations for all the three plankton components studied; the maximum values were 2.48, 0.72, and 4.68 recorded for phytoplankton, tintinnids, and mesozooplankton, respectively.

The lowest diversity index for tintinnids was found at Gangasagar (S_5) during postbloom period. Diversity decreases when a community becomes dominated by one or a few species; therefore, when communities are highly uneven, or there is extreme dominance by one or a few species, their functioning is less resistant to environmental stress (Alimov et al. 2010).

One way ANOVA was performed taking into consideration of the numerical abundance of total tintinnids, phytoplankton, and mesozooplankton with respect to phases of distribution (pre- and postblooming period) and stations as different factors. Significant value was obtained for phases ($F = 3.09$, $p < 0.05$), whereas insignificant for stations ($F = 0.14$, $p < 0.9$) when phytoplankton abundance was considered. Similar results were also obtained for tintinnids (between phases: $F = 4.83$; $P < 0.02$ and between stations: $F = 0.10$; $p < 0.91$) and between mesozooplankton (phase: $F = 10.72$; $p < 0.00$ and Stations: $F = 1.23$; $p < 0.29$).

A multispecies bloom, caused by the centric diatoms, viz. *Coscinodiscus radiatus*, *Chaetoceros lorenzianus* as well as the pennate diatom *Thalassiothrix frauenfeldii* was also recorded in coastal regions of western part of Sundarbans, and similar adverse impact on phytoplankton and microzooplankton was also evident due to occurrence of such bloom. Both number (15–18 species) and cell densities (12.3×10^3 cells l^{-1}–11.4×10^5 cells l^{-1}) of phytoplankton species increased during peak bloom phase exhibiting moderately high species diversity ($H' = 2.86$), richness ($R' = 6.38$), and evenness ($E' = 0.80$). The diatom bloom, existed for a week, had a negative impact on the tintinnid community in terms of drastic changes in species diversity index (1.09–0.004) and population density (582.5×10^3–50×10^3 ind m^{-3}). The bloom is suggested to have been driven by the aquaculture activities and river effluents resulting high nutrient concentration at this region. An attempt has been made to correlate the satellite remote sensing-derived information to the bloom conditions. MODIS-Aqua-derived chlorophyll maps have been interpreted.

4.2 Interaction Between Climate Change and Tintinnid Community

Sundarbans coastal regions and the adjacent Ganges River Estuary (GRE) are potential examples of major wetland complexes situated in the Ganges delta which is tectonically active and geochemically youngest river basin in the world. Sundarbans can be recognized as the most delicate, vulnerable and globally threatened ecosystems residing at the boundary between land and water. These are subjected to driven changes in the structure and functioning of ecosystem most susceptible to the impact of climate change. The possible means by which climate variability affects these wetland ecological processes are many and vary across a broad range of spatiotemporal scales. This may affect either physiology of living organisms (metabolic and reproductive processes), biological environment (predator–prey interactions), or abiotic environment (habitat type and structure).

Ecological response to the climatic variations may be immediate or lagged, and linear or nonlinear or may result from the interactions between climate and other sources of variables. The climate change drives of vulnerability in the coastal region could be referred as catastrophic cyclone and strong winds, erosion, inundation and saltwater intrusion, flood and storm surge, intense rain, extreme temperature, and sea-level rise. The Bay of Bengal has a tropical cyclone frequency of 20 %, and it is observed that the sea level quickly due to strong winds and decreased atmospheric pressure (storm surges) and the waves can reach heights up to 6–7 m when the cyclones reach the coast (Barry and Charley 1987). This cause to submerge the low-elevation delta areas along the coastal fringe and floods the adjacent agricultural fields, destruct the embankments and sometimes cause disasters when the cyclone coincides with the high tides.

In addition, the rising trend in sea level and surface water temperature was also evident at the regional scale and may be a threat to these low-lying coastal regions in the context of erosion, inundation, and seawater intrusion. Nandy and Bandopadhyay (2011) showed that the trend of annual sea-level rise along the Ganges (Hugli) River coast 1.09 mm/year causing inundation variable at regional scale depending on the topography of the region. For SST, a trend of 0.05 °C/year was recorded in Indian Sundarbans wetland by Mishra et al. (2009). The Intergovernmental Panel on Climate Change (IPCC) reviewed that 20 % of species assessed may be at risk of extinction from climate change.

Climate change can be liable to aggravate the present threat to biodiversity in Sundarbans through the following processes: (a) direct mechanism such as habitat loss/destruction for species or reduced resilience in the ecosystem due to pervasive pressure from multiple anthropogenic stressors (such as eutrophication, dredging, excavation of sand from the river bed, decreasing of water quality) and (b) indirectly through its impact on human and their dependence on the living and non-living resources produced by the estuarine ecosystem. The response of expanding human population to the climate change will place considerable pressure on the biodiversity. More logistic consideration is the degradation taking the forms of continuing less in the natural mangrove habitat can influence processes affecting climate change through the release of carbon dioxide to the atmosphere (Van der Werf et al. 2009).

The Ganges River is within watersheds that are stressed to same extent by human activities including development and dam destruction. Climate change will add to magnify risk that is already present through its potential to alter rainfall, temperature, and runoff patterns and disrupt biological communities and severe ecological linkage.

Increased frequency and intensity of tropical cyclones, flood, storm surges, and other extreme catastrophic events are the key drivers of biodiversity patterns including microzooplankton and thus affecting community structure and ecosystem function processes.

Microzooplankton ciliates (including tintinnids) dominate trophic interactions and biogeochemical processes at the base of the marine pelagic food webs. The key role of these small-sized non-commercial microzooplankton is mainly ascribed to their short life cycles and their sensibility to environmental changes (Hays et al. 2005). It is assumed that microzooplankton biomass, species composition, and trophic activities would be responsible to a wide array of changes in bottom-up and top-down controls due to cumulative impact of climate change and have potentially large implications for functioning of marine ecosystems.

Autochthonous key tintinnid species in the coastal regions of Sundarbans wetland as encountered in the present investigation (e.g., *Tintinnopsis beroidea*, *T. minuta,* and *T. lohmani*) are the major controllers of system function in microzooplankton. They exhibit unique capacity to adjust to changing environmental conditions. Again, the given species may be rare and less abundant (such as *Favella ehrenbergii)* but change dramatically in abundance and importance at other times, supporting the idea of biodiversity as "ecosystem insurance" (Yachi and Loreau 1999). Monti et al. (2012) observed substantial changes in tintinnid composition in the Gulf of Trieste over the period of 1986–2010 with the almost complete disappearance of some species and the arrival of some "new entries." There are cascading consequences of these impacts on microzooplankton community, including altered food web dynamics, reduced abundance of habit-forming species, range shifts, and expansion of pathogens (Hoegh-Goldberg and Bruno 2010).

Extreme catastrophic events are key drivers of biodiversity pattern and that the frequency and intensity of such episodes have great implications on species distribution and ecosystem structure. Rapidly changing climates and habitats may increase opportunities for a suite of invasive or alien species to spread at the expense of native species because of their adaptability to disturbance. Recently, Saccà and Giuffrè (2013) highlighted the invasive behavior of the neritic tintinnid *Rhizodomus tagatzi* which can tolerate ample salinity shifts (16–37 p.s.u.) and prefer relatively warm conditions (above 18–29 °C). They had suggested that aquaculture transplants could act as a possible means of dispersion for these conspicuous ciliated protozoa.

The metropolitan megacity, Calcutta, situated about 145 km off from the mouth of the Ganges estuary, would rank 8th with a population of 20.1 million in the context of urban agglomerations in 2025 (based on data UN, 2010). This has also central importance for environmental complexity and related ecosystem services of the adjacent coastal environments of the Ganges River and Sundarbans environments. The city acts as the important driver for socioeconomic development but also potential source of environmental challenge. This can be regarded as a major center of energy used, waste production, and generation of heat-trapping greenhouse gases (GHG) and thus considered local focal points for efforts to reduce GHG emissions. In addition, the huge pollution load can damage downstream coastal zone ecosystem function, and resources including fish and fisheries induce algal bloom (as previously discussed) and feedback to the atmosphere via marine emission. Calcutta is expected to be affected in numerous ways to climate change, and it is essential to identify and implement adaptive strategies by the decision makers to deal with the potential negative impacts of a warming climate (e.g., Scott et al. 2001). Increasing hazards such as floods, storms, coastal erosion and inundation, heat extremes, and air pollution are among the expected impacts, all experienced in the context of multiple other stresses and global changes (Tebaldi et al. 2006).

The impact of climate change in future would be quite severe for the megadeltas in Asia. IPCC has projected that with the rise in temperature and subsequent rise in sea level in the coasts of Asia, including the Indian Sundarbans, will be exposed to increasing risks like coastal erosion. WWF-India has projected rising trend in surface temperature (0.019 °C year^{-1}) for the Sundarbans region and predicted that more than 1.3 million people will be affected by the sea-level rise and permanent submergence of land masses, storm surges, and coastal erosions. The effect of climate change on microzooplankton will have severe adverse impacts on phytoplankton community and density and herbivorous copepods and thereby affecting ecosystem services, such as carbon sequestration, oxygen production, and biogeochemical cycles.

The megadelta provides diverse physiographic characteristics and habitat varieties and thus equipped with high species diversity and endemism which make this delta region priority for conservation. Combined effects of anthropogenic stresses and regional climate change are very much discernible for the sustenance of vulnerable species and ecosystems in Sundarbans and need immediate conservation measures to secure and maintain its ecosystem stability. This is required to maintain "ecological integrity" which is the ability of an ecosystem to support and maintain a community of organisms with species composition, diversity, and functional organization similar to those within natural habitats in the same regions (Parrish et al. 2003).

4.3   Conclusion

The detailed case study elucidated a combined taxonomic and ecological aspects of the tintinnids, a species-rich group of marine protistan microplankton, from the coastal regions of a tropical iconic Sundarbans mangrove ecosystem. This provides some new information on their spatiotemporal distribution and community structure in the context of environmental variables along with the effective role of multiple stressors. Each study site in Sundarbans is characterized by a specific fingerprint of dominant tintinnid assemblages being governed by specific environmental variables, especially water temperature, chl *a,* and salinity followed by inorganic nutrients. In addition, tintinnid diversity is intricately related to complex mechanisms such as various biological factors (predators, resilience of dominant tintinnids, etc.,) as well as to particular physical and hydrological characteristics (water currents, water column stability, wind and wave action, etc.,) of each sampling site. The overall dominance of small-sized tintinnids (<40 μm) at higher concentrations for all study sites in Sundarbans suggests their eutrophic nature and seems to a consequence of the high availability of bacteria and small flagellates. Tintinnid diversity more closely reflects resource diversity (phytoplankton size diversity) than competitive interactions on predations, and abundance and dominance might be straight indicators of ecosystem trophic state. The ubiquitous presence of the agglutinated species *T. beroidea* might be considered as a key species which revealed distinct seasonal variations in oral diameter and length of the lorica mainly linked to water temperature. Differences in the occurrence of less abundant tintinnid species also suggest an adjustment in the tintinnid community, as in case of hyaline species *F. ehrenbergii*.

Further investigations about the trophic relationship between ciliated and other microbial and metazoan components are required for better understanding of their role in the microbial food webs. There is need for further studies on the diversity and abundance of pico- and nano-plankton to get a holistic view of the trophic-level status in Sundarbans, the most vulnerable and productive wetland ecosystem.

4.4 Recommendations

(a) Climate change is causing significant trends of seasonal timing (e.g., generation time, migration pattern, species interactions, animal movement), recognized as phenology, in a wide variety of biota and environment (Durant et al. 2007). This phenomenon can provide sensitive indicators of climate change. Hence, this would be the utmost importance to motivate future research to identify and quantify the year-wise variability for broader understanding of phenology at the microzooplankton species and community level through innovative methodological approaches.

(b) Both impact of climate change and human-induced stresses are responsible to have a profound effect on the structure and productivity of marine systems, and they should be considered as primary and potential drivers of global environmental changes. Climate change, creating frequent catastrophic events, may enhance the ability of certain invasive or allochthonous species to invade coastal region of Sundarbans and more chances to survive, reproduce, and compete with native species. The unpredictability and strong impacts of these invasions demand long-term coherent management strategies to control substantial negative impact on native biota, economic value, or human health.

(c) Recently, the concept of ecosystem-based adaptation (EBA) has been developed that uses ecosystem services as part of an overall adaptation strategy to help people to adapt to the adverse effects of climate change (Andrade et al. 2011). EBA is mainly concerned for sustainable management for both target and non-target species by preserving or restoring habitat quality to maintain ecosystem services (Rosenberg and Mcleod 2005). For Sundarbans wetland ecosystem, EBA might be applied which would play a crucial role in climate change adaptation where about 6.5 lakh inhabitants are quite dependent directly or indirectly on natural resources mainly on mangroves and mangrove-related products (such as wood and honey) as well as marine fisheries.

References

Andrade, C. (coord), Antunes, C., Marques. F. et al. (2011). Criação e implementação de um sistema de monitorização no litoral abrangido pela área de jurisdição da Administração da Região Hidrográfica do Tejo, I.P.. 3° Progress Report (unpublished) p.113.

Alimov, A. F. (2010). Changes in the structure of animal communities accompanying eutrophication and pollution of aquatic ecosystems. *Doklady Biological Sciences, 433,* 249–251.

Barría de Cao, M. S., Beight, M., & Piccolo, M. C. (2005). Temporal variability of diversity and biomass of tintinnids (Ciliophora) in a southeastern Atlantic temperate estuary. *Journal of Plankton Research 27*(11), 1103–1111.

Barría de Cao, M. S., Pettigrosso, R. E., & Popovich, C. (1997). Planktonic ciliate during a diatom bloom in Bahia Blanca Estuary, Argentina, II. Tintinnids. *Obelia 23,* 21–31.

Barry, R. G., & Charley, R. J. (1987). *Atmosphere weather and climate.* London: Methuen.

Devassy, V. P., & Nair, S. R. (1987). Discoloration of water and its effect on fisheries along the Goa coast. *Mahasagar Bulletin of the National Institute of Oceanography, 20,* 121–128.

Diersing, N. (2009). *Phytoplankton blooms: The basics* (2 pp). Florida Keys National Marine Sanctuary, Key West, Florida, USA. Available at http://floridakeys.noaa.gov/pdfs/wqpb.pdf.

Dolan, J. R., Claustre, H., Carlotti, F., Plounevez, S., & Moutin, T. (2002). Microzooplankton diversity: Relationship of tintinnid ciliate with resources, competitors and predators from the Atlantic coast of Morocco to the Eastern Mediterranean. *Deep Sea Research Part I: Oceanographic Research Papers, 49,* 1217–1232.

Duguay, L. E., Monteleone, D. M., & Quaglietta, C.E. (1989). Abundance and distribution of zooplankton and ichthyoplankton in Great South Bay, New York during the brown tide outbreaks of 1985 and 1986. In E. M. Cosper, E. J. Carpenter & V. M. Bricelj (Eds.), *Coastal and Estuarine Studies* (Vol. 35, pp. 599–623). Novel phytoplankton blooms: Causes and impacts of recurrent brown tides and other unusual blooms. New York: Springer; *Ecology 22,* 20–30.

Durant, J. M., Hjermann, D. O., Ottersen, G., Stenseth, N. C. (2007). Climate and the match or mismatch between predator requirements and resource availability. *Climate Research, 33,* 271–283.

Falkowski, P., Barber, R., Smetacek, V. (1998). Biogeochemical controls and feedbacks on ocean primary production. *Science, 281,* 200–206.

Hays, G. C., Richardson, A. J., & Robinson, C. (2005). Climate change and marine plankton. *Trends in Ecology and Evolution, 20*(6), 337–344.

Hoegh-Goldberg, O., & Bruno, J. (2010). The impact of climate change on the World's marine ecosystems. *Science, 328,* 1523–1528.

Hu, H., Zhang, J., & Chen, W. (2011). Competition of bloom-forming marine phytoplankton at low nutrient concentrations. *Journal of Environmental Sciences, 23*(40), 656–663.

Jayaraman, R., & Seshappa, G. (1957). Phosphorus cycle in the sea with particular reference to tropical inshore waters. *Proceedings of the Indian Academy of Sciences, 46,* 110–125.

Li, X. Y., Zhang, S. X., & Zhang, J. (2009). Mesoproterozoic calymmian tintinnids from central China. *The Open Paleontol Journal, 2,* 10–13.

Lynn, D., & Montganes, D. J. (1991). Global production of heterotrophic marine planktonic ciliates. In P. C. Reid, C. M. Turley & P. H. Burkill (Eds.), *Protozoa and their marine process* (Vol. G 25, pp. 281–307). NATO ASI series. Berlin: Springer.

Mishra, P. K., Mishra, S., Bisht, S. C., Selvakumar, G., Kundu, S., Bisht, J. K., et al. (2009). Isolation, molecular characterization and growth-promotion activities of a cold tolerant bacterium *Pseudomonas* sp. NARs9 (MTCC9002) from the Indian Himalayas. *Biological Research, 42,* 305–313.

Monaldo, F. M., Sikora, T. D., Babin, S. M., & Sterner, R. E. (1997). Satellite imagery of sea surface temperature cooling in the wake of Hurricane Edouard, (1996). *Monthly Weather Review, 125,* 2716–2721.

Monti, M., Minocci, M., Milani, L., & Fonda Umani, S. (2012). Seasonal and interannual dynamics of microzooplankton abundances in the Gulf of Trieste (northern Adriatic Sea, Italy). *Estuarine. Coastal and Shelf Science 115,* 149–157.

Nandy, S., & Bandopadhyay, S. (2011). Trend of sea level change in the Hugli estuary India. *Indian Journal of Geo-Marine Sciences, 40*(6), 802–812.

Naz, T., Burhan, Z., Munir, S., & Siddiqui, P. J. A. (2012). Taxonomy and seasonal distribution of *Pseudonitzschia* species (Bacillariophyceae) from the coastal water of Pakistan. *Pakistan Journal of Botany, 44*(4), 1467–1473.

Parrish, J. D., Braun, D. P., & Unnasch, R. S. (2003). "Are we conserving what we say we are?" Measuring ecological integrity within protected areas. *BioScience, 53*, 851–860.

Pielou, E. C. (1969). *An introduction to mathematical ecology*. New York: Wiley.

Rosenberg, A. A, & McLeod, K. L. (2005). Implementing ecosystembased approaches to management for the conservation of ecosystem services. *Marine Ecology–Progress Series, 300*, 270–74.

Round, F. E., Crawford, R. M., & Mann, D. G. (1990). *The diatoms: biology and morphology of the genera* (pp. 1–747). Cambridge: Cambridge University Press.

Saccà, A., & Giuffrè, G. (2013). Biogeography and ecology of *Rhizodomus tagatzi*, a presumptive invasive tintinnid ciliate. *Journal of Plankton Research, 35*(4), 894–906.

Sachithanandam, V., Mohan, P.M., Karthik, R., Sai Elangovan, S., & Padmavathi, G. (2013). Climate changes influence the phytoplankton bloom (Prymnesiophyceae: *Phaeocystis* spp.) in North Andaman coastal region. *Indian Journal of Geo-Marine Sciences 42*(1), 58–66.

Santhanam, R. (1976). PhD thesis (p. 101). Annamalai University, Chidambaram, India.

Sarangi, R. K., Chauhan, P., & Nayak, S. R. (2005). Inter-annual variability of phytoplankton blooms in the northern Arabian Sea during winter monsoon period (February–March) using IRS-P4 OCM data. *Indian Journal of Marine Sciences, 34*(2), 163–173.

Sarkar, S. K., Saha, M., Takada, H., Bhattacharya, A., Mishra, P., & Bhattacharya, B. (2007). Water quality management in the lower stretch of the River Ganges, East Coast of India: An approach through environmental education. *Journal of Cleaner Production, 15*(16), 1459–1467.

Satpathy, K. K., Mohanty, A. K., Sahu, G., Natesan, U., Venkatesan, R., & Prasad, M. V. R. (2007). On the occurrence of *Trichodesminum erythraeum* (Ehr.) bloom in the coastal waters of Kalpakkam, east coast of India. *Indian Journal of Science and Technology, 1*, 1–11.

Saunders, R. D., & Glenn, D. A. (1969). Diatoms. *Memories of Hourglass Cruises, 1*, 119.

Scott, J. M., Davis, F. W., Mcghie, R. G., Wright, R. G., Groves, C., & Estes, J. (2001). Nature reserves: Do they capture the full range of America's biological diversity? *Ecological Applications 11*(4), 999–1007.

Shetty, H. P. C., & Saha, S. B. (1971). On the significance of the occurrence of the blooms of the diatom *Hemidiscus hardmannianus* (Greville) Mann in relation to *Hilsa* fishery in Bengal. *Current Science, 15*, 410–411.

Smayda, T. J. (1997). Bloom dynamics: Physiology, behavior, trophic effects. What is a bloom? A commentary. *Limnology and oceanography 42*(5, part 2), 1132–1136.

Subramanian, A., & Purushothaman, A. (1985). Mass mortality of fish and invertebrates associated with a bloom of *Hemidiscus hardmannianus* (Bacillariophyceae) in Parangipettai (Southern India). *Limnololgy and Oceanography, 30*(4), 910–911.

Takahashi, M., Seibert, D. L., & Thomas, W. H. (1977). Occasional blooms of phytoplankton during summer in Sannich Inlet. *Deep Sea Research, 24*, 775–780.

Tebaldi, C., Hayhoe, K., Arblater, J. M., & Meehl, G. A. (2006). Going to the extremes: An inter-comparison of model-simulated historical and future changes in extreme events. *Climate Change, 79*, 185–211.

Van der Werf, G. R., Morton, D. C., DeFries, R. S., Olivier, J. G. J., Kasibhatla, P. S., Jackson, R. B., Collatz, G. J., & Randerson, J. T. (2009). CO_2 emissions from forest loss. *Nature Geoscience 2*, 737–738.

Yachi, S., & Loreau, M. (1999). Biodiversity and ecosystem productivity in a fluctuating environment: The insurance hypothesis. *Proceedings of the National Academy of Sciences, USA. 96*(4), 1463–1468.

Glossary

Abundance Abundance is an ecological concept referring to the relative representation of a species in a particular ecosystem. It is usually measured as the large number of individuals found per sample.

Agglomerate A mass or clump of things gathered together by means of any physical interaction agglomerate; an unmethodical assemblage; a cluster.

Bioindicator Bioindicator is an organism which acts as a tool for monitoring the health or measuring the changes that occur in the surroundings in an ecosystem. They produce certain molecular signals in response to changes in their environmental conditions.

Biomass Biomass is the mass of living biological organisms in a given area or ecosystem at a given time.

Biomonitor Biomonitor can be defined as an organism that provides quantitative information on the quality of the environment around it.

Choreotrichous The choreotrichs are a group of small, marine ciliates. Their name reflects the impression that they appear to dance ('choreo' as in choreography).

Community The term 'community' is used to designate an assemblage of ecologically similar, coexisting organisms in an ecosystem or a particular environment.

Core species The species that exist in an annual cycle irrespective of environmental fluctuations.

Correlation matrix Correlation refers to any of a broad class of statistical relationships involving dependence. Correlations are useful because they can indicate a predictive relationship that can be exploited in practice. It refers to any departure of two or more random variables from independence, but technically it refers to any of several more specialized types of relationships between mean values. The Pearson correlation coefficient indicates the strength of a linear relationship

© The Author(s) 2015
S.K. Sarkar, *Loricate Ciliate Tintinnids in a Tropical Mangrove Wetland*,
SpringerBriefs in Environmental Science, DOI 10.1007/978-3-319-12793-4

between two variables, but its value generally does not completely characterize their relationship.

Dendrogram A dendrogram is a tree-structured graph used in heat maps to visualize the result of a hierarchical clustering calculation. The result of a clustering is presented either as the distance or the similarity between the clustered rows or columns depending on the selected distance measure. It is a visual representation of the spot correlation data. The individual spots are arranged along the bottom of the dendrogram and referred to as leaf nodes and directly represented as a nested list where each component corresponds to a branch of the tree.

Diatoms Diatoms are a major group of algae, and are among the most common types of phytoplanktons. Diatom communities are a popular tool for monitoring environmental conditions, past and present, and are commonly used in studies of water quality.

Diversity Degree of variation of life forms within a given ecosystem.

Diversity index A diversity index is a statistical measure of species diversity in a community. Diversity indices provide more information about community composition than simply species richness (i.e., the number of species present); they also take the relative abundances of different species into account. It gives the information about rarity and commonness of species in a community. The ability to quantify diversity in this way is an important tool for biologists trying to understand community structure.

Ecosystem diversity The variety of habitats, living communities, and ecological processes in an area and its surroundings is called ecosystem diversity.

Excystment The term Excystment means one that has been enclosed in or as if in a cyst. The other meaning is one that has the form of enclosed cyst. The migration and excystment of larvae sometimes cause fever and pain brought upon by the host inflammatory response.

Indicator species An indicator species are the biological species whose presence, absence, or relative well-being in a given environment is a sign of the overall health of its ecosystem. Indicator species can be among the most sensitive species in a region, and sometimes act as an early warning to monitoring biologists. By monitoring the condition and behavior of an indicator species, scientists can determine how changes in the environment are likely to affect other species that are more difficult to study.

Invasive species (synonyms: alien, exotic, non-native, allochthonous) Species that are introduced to an area outside their natural range have been identified as one of the major threats to the maintenance of biodiversity and functioning of marine ecosystems. Their introduction may be accidental or intentional and the four main paths of invasion are: aquaculture transfer, ships' ballast water, hull fouling, and the establishment of new connections between sea via canals.

K-Dominance curve The K-dominance curve is a powerful tool for measuring abundance trends in communities over time. K-dominance curves are the cumulative ranked abundance against a log species rank. The logic behind the use of these curves as indicators is that only the subset of species that can tolerate perturbation will thrive and the rest will decline or disappear. Thus, the steepest and most elevated curve shows the lowest diversity and the most perturbed system state. This metric has wide application for measuring changes in relationships over time or comparing species-based areas and could potentially be applied to track species assemblages that have been identified as diagnostic of habitat types.

Lorica A lorica is a shell-like protective outer covering, often reinforced with sand grains and other particles that some protozoans and loricifera metazoans secrete. Usually it is tubular or conical in shape, with a loose case that is closed at one end.

Mesozooplankton Mesozooplanktons are a group of heterotrophic and mixotrophic planktonic organisms in the size range 0.2–20 mm. They play a crucial role in determining the fate of primary production, the composition and sedimentation rate of sinking particles, and thus the flux of organic matter to the deep ocean. Thus they maintain the marine food chain.

Microzooplankton Microzooplankton are a group of heterotrophic and mixotrophic planktonic organisms between 20 and 200 μm in size. They occupy the key position in the marine food chain transferring carbon and energy to organisms of the higher trophic level. Flagellates, ciliates, radiolarians, etc., are included in this group.

Panmictic Random mating of individuals within a population, the breeding individuals showing no tendency to choose partners with particular traits.

Phenology Study of cyclic and seasonal natural phenomena, especially in relation to climate and plant and animal life.

Phylogenetics Phylogenetics is the study of evolutionary relationships among groups of organisms (e.g., species, populations), which are discovered through molecular sequencing data and morphological data matrices.

Principal component analysis (PCA) Principal component analysis (PCA) is a statistical procedure that uses an orthogonal transformation to convert a set of observations of possibly correlated variables into a set of values of linearly uncorrelated variables called principal components. It the simplest of the true eigenvector-based multivariate analyses. Often, its operation can be thought of as revealing the internal structure of the data in a way that best explains the variance in the data.

Production rate The number of units of output that can be produced during a given period.

Satellite species Which are sparse and occur at only a few sites. Satellite species interact very little and are not predictable or slightly predictable to the host environment and therefore, they play little or no role in structuring the infracommunity.

Succession Ecological succession is the observed process of change in the species structure of an ecological community over time.

Tintinnid Tintinnid are ciliates of the choreotrich taxon Tintinnida, distinguished by vase-shaped shells called loricae, which are mostly protein but may incorporate minute pieces of minerals.

Useful websites for further information on Microzooplankton:
http://www.biolo.bg.fcen.uba.ar/planing.htm,
http://www.microzooplankton.uconn.edu/people.html.

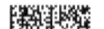